FUTURE
SUPERHUMAN

Elise Bohan is a Senior Research Scholar at the University of Oxford's Future of Humanity Institute (FHI). She holds a PhD in evolutionary macrohistory (Big History) and wrote the world's first book-length history of transhumanism as a doctoral student. At FHI, she is part of a cohort of scholars who are dedicated to understanding, and tackling, humanity's most pressing problems.

'There's much doom and gloom about humanity's future, understandably so at a time of climate change and large-scale environmental collapse. But there's another side to what's coming, and it won't all be bad. Elise Bohan is your travel guide to the future of human minds and bodies. Enjoy the trip – your guide is as sharp, savvy, lively and entertaining as you could ever want.'

RUSSELL BLACKFORD, author of
At the Dawn of a Great Transition

'One of the most entertaining, fascinating, and thought-provoking books I've read in a long time! *Future Superhuman* provides a breathtakingly original, broad and optimistic view of a transhuman future. Elise Bohan's fresh and unique voice comes through on every page: bold, fearless, fun and relentless on the absurdities of the human condition. She could well be the next big non-fiction star, at home in the same constellation as Yuval Noah Harari, Carl Sagan or Elizabeth Kolbert.'

ROB BROOKS, Scientia Professor of Evolution at UNSW, author of *Artificial Intimacy*

'A brilliant, engaging and edgy introduction to transhumanism – the idea that in coming centuries we humans will take charge of our own evolution and transform ourselves into new, artificially enhanced beings.'

DAVID CHRISTIAN, Distinguished Professor of History at Macquarie University, author of *Origin Story*

FUTURE
SUPERHUMAN

OUR TRANSHUMAN LIVES
IN A MAKE-OR-BREAK CENTURY

ELISE BOHAN

NEWSOUTH

A NewSouth book

Published by
NewSouth Publishing
University of New South Wales Press Ltd
University of New South Wales
Sydney NSW 2052
AUSTRALIA
https://unsw.press

© Elise Bohan 2022
First published 2022

10 9 8 7 6 5 4 3 2 1

 A catalogue record for this book is available from the National Library of Australia

ISBN: 9781742236759 (paperback)
          9781742238357 (ebook)
          9781742239255 (ePDF)

Internal design Josephine Pajor-Markus
Cover design Arielle Nguyen
Cover image metamorworks/Shutterstock

All reasonable efforts were taken to obtain permission to use copyright material reproduced in this book, but in some cases copyright could not be traced. The author welcomes information in this regard.

All currency is in US dollars unless noted otherwise.

DEDICATED TO

Antonia and Antonio Torchia (Nina & Tony)
For building a better future

The enemy that has no name is not a nation, an organization or a religion. It is not a corporation or an industry. It is not an economic system or an ideology. It is a way of living on the earth that evolved, and if we are to change it, we must take evolution from autopilot and into our own hands. We must come together to create the future we wish to inhabit.

— BRET WEINSTEIN

Having raised humanity above the beastly level of survival struggles, we will now aim to upgrade humans into gods, and turn Homo sapiens *into* Homo deus.

— YUVAL NOAH HARARI

Marty McFly: *Doc, what if we don't succeed?*

Doc Brown: *We MUST succeed!*

— BACK TO THE FUTURE PART II

CONTENTS

PREFACE

Future Superhuman is a love letter to humanity, though an unconventional one. Like most people, I want the world to be the best possible place, for suffering to be minimal, opportunities vast, love abundant, and kindness, compassion and a deep sense of our common humanity foregrounded as we tackle complex problems as a united global tribe.

But I believe in showing love through honesty, which often includes challenging people when you think they might have blind spots and encouraging them to probe the areas where they're most afraid to look. That is how I feel the most radiant forms of growth are kindled – in individuals, and the species.

This tough love approach will not be for everyone. In my personal relationships I deliver it with great care. Deep intimacy and growth take time and eventually you know when to gently encourage and when to push. But I do not have the luxury of time with my readers. You are each so different. I don't know your individual quirks, trigger points, fears and backstories and I can't anticipate them all or always be sensitive to each of them. That doesn't mean I think your feelings are irrelevant. It simply means I trust you to sit with, and process, them in your own ways and in your own time.

While I would love to see other writers take you by the hand more gently (as I attempted to do in earlier drafts, without success), the subject matter doesn't lend itself to a soft approach. So, I made the conscious choice to get to the heart of the things that I think matter most and to do it swiftly and directly. The point is not to dismiss the many topics and facets of topics that recede into the background. It's to foreground ideas that prompt you to think outside the frameworks you're most comfortable thinking within.

The bravest among you will rise to that challenge and question whether, beneath some of your discomfort, lies fear. From there, you may consider whether it's fear, rather than righteousness, that is triggering the impulse to dismiss, or deride. Others will declare without a moment's pause that I am wrong about many ideas, simply because I've presented them in a way that challenges what they happen to presently believe. To those readers, I encourage you to consider whether there might be some truth or validity to both perspectives.

The difficulty of a project like this speaks to the inherent complexity of the world we live in. By zooming out and taking a broader view you necessarily lose detail. By foregrounding confronting ethical and intellectual conundrums you lose a degree of comfort, camaraderie and reassurance. Do not mistake either of those choices for a naivety of the existence, or value, of the alternatives.

I have made the trade-offs I have because my personality compels me to. For better or worse, I have the kind of mind that is obsessed with what lies hidden under the rug. I want to start conversations at the point where they usually peter out. That's because I think our conversations usually wrap up just as they're starting to get interesting (and consequently challenging and risky). At those points it becomes easier to say, 'not today, I'll let that lie'.

The trouble is, if everybody says, 'not today, I'll let that lie' the most important conversations in our society will necessarily be blinkered. It seems to me that the biggest risks and the greatest opportunities for our species lie in the space beyond those blinkers. There is no easy way to bring topics from 'the beyond' into our field of vision. I wish I could have done so in a way that felt bubbly, safe, affirming and easily digestible. But watering down difficult things – or focusing on humanising the conundrums, rather than on the conundrums themselves – will, I fear, not serve us best in the long run.

That doesn't mean the human challenges are lost on me. Or that I'm not plagued by uncertainty when it comes to the nitty-gritty of lots of these ideas. Writing about the future is hard and getting it right in every respect is impossible. But thinking deeply, cobbling together narratives that show possibilities in a new light, encouraging others to contemplate and debate them and to draw their own conclusions – as I encourage you all to do here – is a profoundly valuable exercise.

It is also the kind of exercise that's part of a long and fruitful intellectual tradition. Future possibilities were richly explored by early- to mid-20th century scientists and science communicators, like JBS Haldane, JD Bernal, HG Wells, Julian Huxley and Pierre Teilhard de Chardin – and later by Arthur C Clarke, Alvin Toffler, Carl Sagan and others. Haldane delivered a famous lecture to the Heretics society at Cambridge University in 1923, which was published later that year as a treatise called *Daedalus: or, Science and the Future*. The book was an astonishingly prescient futurist text. In the introduction, Haldane wagered that his work 'will be criticized for its undue and unpleasant emphasis on certain topics'. But he affirmed that 'this is necessary if people are to be induced to think about them, and it is the whole business of a university teacher to induce people to think'.[1]

The novelist and non-fiction author HG Wells also looked ahead with eerie prescience. Among other things, he predicted the emergence of heavier-than-air flying machines and foresaw 'the possibility of a world-wide network being woven between all men about the earth'.[2] But he naturally got plenty of things wrong. In later life, he noted that his friends were apt to point out every weakness, limitation and error in retrospect – we might call this a 'gotcha' moment today. But reflecting on his early writings, an unphased Wells remarked, 'I look back upon them, completely unabashed. Somebody had to break the ice. Somebody had to try out such summaries on the general mind. My reply to the superior critic has always been – forgive me – "Damn you, do it better."'[3]

I respect your right to disagree with any part of this book and I am all ears to solid counterarguments and counterevidence. I am eager to hear both, for my world view is still evolving – and I know that at every juncture of this book there is more to the story. Yet I still think there is value to putting down your considered thoughts in the moment, and to foregrounding many of the tricky ideas that we find hardest to ponder and discuss. In this book, that comes at the expense of giving airtime to many of the ideas (or specific takes on ideas) that we discuss routinely.

It is my deepest hope that this act of intellectual exploration will not be viewed as a dismissal of all that is rich, complex and beautiful about being human. Surely that beauty can withstand an examination of our flaws and limitations – even when the argument is that those flaws, left unaltered, will lead us to the brink of extinction. We can be beautiful, and tragically flawed, at the same time. I hope I will not be perceived by any of my fellow humans as somehow *less* than human for suggesting this, or thought of as a 'baddie' or a 'villain.' That is a hard possibility to reconcile with the positive spirit in which this book was written.

As a historian I cannot help wondering what kind of source this will make when people look back on the 2020s. If nothing else, it is an honest account of a mind attempting to process many of the biggest conundrums of our time. This has not been an easy task. But it was done in the hope that humanity will finally look the biggest challenges we're facing in the eye, come to terms with them, and prepare.

INTRODUCTION

THE BIG PICTURE AND THE HUMAN STORY

Humanity looks to me like a magnificent beginning but not the last word.

Freeman Dyson, quoted in *Great Mambo Chicken and the Transhuman Condition*

Surgical scrubs. Tense faces. A woman was lying on a theatre table on a rainy Thursday while an obstetrician's hands reached in and scooped something out of her midsection. For the obstetrician it was just another day, just another scheduled c-section. The baby had been doing somersaults and managed to wrap the umbilical cord around its neck – a playful trick that backfired, rendering a natural birth very risky. Without medical intervention the infant might have died, or suffered brain damage, as it made its way through the birth canal.

'Thank God for Norman Blumenthal', my dad says every time he thinks back to that day. That's the name of the ob-gyn who delivered me. A hundred years ago there's a good chance I'd have been snuffed out before I could begin living, and my mum's life might have also been in danger. A hundred years before *that*, most doctors hadn't twigged that it was crucial to disinfect their hands between performing autopsies and delivering babies, and many of the resulting infections were fatal. But in 1990, with the aid of modern medicine, my delivery was routine and unremarkable. A triumph of science and technology over nature, reflecting our age-old drive to redefine the limits of self and world.

After leaving Baulkham Hills hospital on the outskirts of Sydney, my parents brought me back to a simple suburban home. The most technologically sophisticated items they owned were a small TV with wood panelling and rabbit ears, a VCR, two second-hand cars, and a white landline telephone with a curly grey cord. They didn't own a fax machine, a mobile phone or a computer, and they were yet to utter, let alone comprehend the significance of the phrase, 'the internet'. In their mind's eye, my future would unfold in a world much like their present.

Fast-forward thirty years and at a glance it might seem like many things in the world have stayed the same. People still live in houses and apartments, go to work, shop, cook, eat, sleep and have children. Nobody's jet-setting off to the moon for a holiday, and we don't have X-ray vision glasses, or flying cars. What's really happened is much more profound, but harder to spot, because it's not flashy gadgets and consumer products that are driving the biggest changes in our world. It's the underlying evolution of information technology.

Over the past thirty years, our minds, and the collective intelligence of our species, have begun to merge with the architecture of a new technological ecosystem, powered by modern computers, servers, undersea cables and satellites. Somewhere far away from most of us, in a mythical-sounding place called Silicon Valley, we hear that nerds are building robots and algorithms that can walk, talk, jump, fly, see, learn, drive, write, and kick your butt at playing chess, Go and poker. But so what? There's work tomorrow, and the day after that, bills to pay, and kids to raise. Besides, people have always been inventing whimsical new creations and overpromising about a space-age future that never seems to come to fruition.

That's where our minds play tricks on us. We're in the future already. It just lacks the aesthetic of a Hollywood blockbuster set. Nobody expected things to change as fast or as profoundly as they did in the 20th century. My grandparents were born into a world of ice boxes and horse-drawn carts, and two of them are still here in a world of smartphones, an international space station, self-driving cars, and a global, digital brain. We should not expect the disruption to stop here.

Although we don't often recognise it, the 21st century is a *transhuman era*, where everything that currently makes us human, from our brains and bodies, to our values and ways of life, is poised to be transformed or superseded. In our lifetime, we could merge with forms of artificial intelligence that are radically smarter than us, rewrite our biology to conquer ageing, disease and involuntary death, leave behind the crudest and cruellest vestiges of our evolutionary programming, and embrace a new mode of being that is so much more than human that we would have to define it as posthuman. In its best incarnations, we might call this kind of future superhuman.

Welcome to the transhuman era!

This is a book about what it's like to live in a make-or-break century. An era when everything about life as we know it could change rapidly, and radically. These days it's common for academics, journalists and politicians to invoke the truism that we're living in a time of astonishingly rapid change. As true as that statement is, I think the deep implications are often lost on us, as we nod along and visualise the most topical forms of upheaval: leaders suddenly being ousted from their positions, well-known people being deplatformed or cancelled, new gadgets and social media platforms taking the world by storm, and the speed at which digital memes now spread, forcing social norms and narratives to pivot under the pressure of a cascade of likes and shares.

These phenomena are all noteworthy. But they're also symptoms of something bigger, and that something bigger is the focus of this book. What's on the line in the 21st century is greater than our social, cultural, political and institutional norms, our familiar ways of life, our privacy, or our ability to connect, empathise and think deeply about complex issues. It's the long-term survival and flourishing of intelligent life in every guise: human, non-human and superhuman.

There are three main arguments in this book:

1   We're already living in a transhuman era.
2   We need to stay on this transhuman trajectory if we want the future to be bright and sustainable (the big picture).
3   These transition times are probably going to be rough to live through, in spite of the potential perks (the human story).

That first point might not sound very intuitive. After all, humans are flexible creatures and mothers of invention and we've always utilised tools and technologies to extend our reach over the natural world – from flint axes and taming fire; to farming, inventing language and writing; and building civilisations with large populations and complex divisions of labour. With that history in mind, it's reasonable to argue, as many do, that we've always been transhuman.[1]

But *how* transhuman? The speed and degree of change we're experiencing today is novel. We're not transforming from hunter-gatherers to farmers, or from farmers to city-dwelling industrialists and factory workers; we're transitioning from a purely biological species that exerts influence over the natural world using simple tools, and conceptual systems like language, to a much deeper kind of techno-human hybrid that is on the cusp of becoming something categorically new.

That's why it's helpful to think of the present as a characteristically transhuman era. As someone living through this unique historical moment, you're poised to encounter some of the most exciting opportunities, and some of the most profound perils, that our species has ever faced. From genomic sequencing and gene therapy, to designer babies, robot workers, sexbots, and symbiotic connections with smart AI systems that know us better than we know ourselves, human life as we know it is set to change as never before. That's not an easy prospect to get your head around.

The future is stranger than you think

'So you're saying I'm going to become a ... a ... *machine?*' My dad looks at me sceptically, brow creased with mounting stress and a twinge of fear. After a decade's worth of dinner table chats on the subject, he still gets uncomfortable when I ask him to imagine smart technology proliferating, recombining and changing human lifeways at an accelerating rate. He worries that radically transforming humanity would diminish its beauty, and imagines us devolving into cold, inhuman, mechanistic computers, devoid of rich experiences and individuality. That's not where I think we're headed, but it can be hard to know how to respond because there isn't a handy soundbite that sums up the whole story: from here to posthumanity.

There's a lot to unpack in a book as broad and bold as this one. What makes a rapid leap to a superhuman future possible when we've been human for hundreds of thousands of years? Out-evolving humanity also sounds like a scary prospect. Shouldn't we try and stop this rupture in the fabric of our reality? And won't every advance in our technological capabilities come with major attendant risks? I'm going to ask you to sit tight and trust that your burning questions, and understandable scepticism, will be addressed as the story progresses.

What you need to know up front is why I'm writing about this in the first place. I promise I'm not one of those tin-foil-hat-wearing cranks, I research these ideas for a living. My PhD thesis was the first book-length history of transhumanist ideas, movements and technologies that focus on enhancing human capabilities and upgrading our species to a more-than-human state. We'll explore those ideas and their influence on our world in chapter two.

By the time this book is published, I'll have taken up a position at the University of Oxford, at a research hub called the Future of Humanity Institute (FHI). I'm joining a team of scholars who get paid to think about the biggest issues facing our species. These include existential risks (the kind that could put an end to human civilisation and prevent anything like it ever emerging on this planet again) and how to maximise our chances of seeding a positive world for the trillions of possible future beings whose moment in the sun, or dance around another star, is yet to come.

I care a lot about the long-term prospects for intelligent life. But right now, my main concern is what happens to humanity in the next hundred years. There will be no long-term future to safeguard if everything unravels in the meantime. If *Homo sapiens* goes down in history as the smartest and most technologically advanced species to walk the Earth, it means we likely get wiped out in the 21st century, or not long after. The destructive potential of our advanced and emerging technologies is too high, and compounding too fast, for us to safely sit on them for many more millennia.

In this transhuman era of escalating promise and peril, humanity's most pressing task is to prevent major setbacks that could derail our civilisations this century – like runaway global warming, major pandemics, nuclear war, and the rise of forms of artificial intelligence that develop motives and values incompatible with ours. To pull that off, we will need more intelligence, more technology, and more-than-human powers.

From smart fridges to *The Matrix*?

Of course, we're all hankering to know what the future looks, sounds and smells like. But I'm afraid it's impossible to transport you to a realistic posthuman world through even the most vivid storytelling. The point at which we are no longer human is like an event horizon: it's a rupture in the fabric of our subjective reality. Our ape-brains weren't built with the capacity to imagine what it would be like to be millions of times smarter, to think millions of times faster, to have radically more bandwidth, or totally different motivations, preferences and desires than those evolution has baked into our bodies and minds.

I could talk about self-proclaimed cyborgs, like the colour-blind artist Neil Harbisson who had an antenna implanted in his skull, enabling him to perceive colour through soundwaves. Or I could wax ecstatic about bio-hackers with radio frequency identification (RFID) chips implanted in their wrists, which allow them to make digital payments through their skin. These stories are engaging, and they hint at the possibilities for human enhancement that wearable and implantable technologies could open up. But in their current manifestations, add-ons like this are mostly red herrings.

So are most of the 'day in the life' stories that writers concoct to give you a sense of the near-future changes you might experience on the road to a more radically enhanced future. Why would I want a smart fridge that knows when I've used up the cream cheese and automatically reorders it? Maybe I only buy cream cheese a few times a year. I don't want to scurry around correcting the admin errors of my uppity Kelvinator. Nor do I want an AI system to talk to my coffee machine and make sure there's a brew waiting for me at 8 am every morning when I flop out of bed. There's too much room for error. Maybe I'll roll over and go back to sleep, get distracted by a phone call, or lie there reading a sensational news story (let's hope not, but it happens).

I don't particularly want to spend my evenings custom-designing tomorrow's outfit, trying it on in a virtual fitting room and having a drone deliver it to my doorstep overnight. In fact, I don't want to live in a world where the sky is thick with delivery drones at all. I might be spectacularly wrong about this, but the widespread use of aerial delivery drones strikes me as a really bad idea and one that won't stick. Drones are intimidating. They can be used to stalk, surveil and assassinate people. They come with a massive yuck factor and people will mistrust and dislike them. Friendly delivery robots, maybe. Flying black cameras, no.

The whole point of a well-designed future, enhanced by technology, is that we don't notice the upgrades most of the time. Like wi-fi, they're integrated into our surroundings, running through us and our devices, connecting mind with mind, and mind with the cloud. We're painfully aware of our computers when they're running slow, and routers when the internet drops out. But when they're operating smoothly, they feel like extensions of our bodies and minds as we tap away seamlessly on our devices. They're just something that *is* – and that's the real story!

We're sleepwalking into deeper immersion with technologies that *just are* a fundamental part of us now. Becoming more-than-human is the logical next step – which is obvious when we zoom out a little. Most of the biggest changes in human history have taken place in the past 250 years, or *in the most recent 0.1 per cent of our tenure on Earth*. I'm talking germ theory, electromagnetism, quantum theory and relativity, the Darwinian revolution and the neo-Darwinian synthesis, the invention of antibiotics and vaccines, human-induced global warming, planes, trains, automobiles, spacecraft, rockets, nuclear weapons, factory farming, telephones, television, computers, the internet – and the explosion of change that the web has enabled.

In an evolutionary blink of an eye, *Homo sapiens* has gone from being just another great ape to the most powerful species on the planet, and we're continuing to transform the Earth's systems, our own habitats, and our cultures, minds and bodies at an accelerating rate. Entering a transhuman era is not a rupture in evolution any more than the emergence of DNA, multicellular life, brains or humans. It's something new from something old that extends evolutionary potential into new domains. A posthuman future will be the next something new, if nothing goes too badly wrong in the meantime. And it's up to us to make sure that it doesn't.

The big picture

This book tells two stories. The first is the big-picture one, where we look at the present from above and see an impending fork in the evolutionary road. That fork will either lead us to a superhuman future, or extinction. Then there's the human story, where we explore the experience of this transition from within and try to figure out how to navigate our journey between the human world that is on its way out, and the posthuman world that is on its way in.

Here's a helpful way of thinking about the big picture. Humanity is on a ship that's sailing directly towards an iceberg. We have time to change course, but we're a little preoccupied because we're engaged in a complex juggling act.

First two balls, then three, then four. As time wears on, the balls are supplanted by live grenades that can detonate on impact. Quick, catch the next one – it's labelled 'nukes'. And the next – 'pandemics'. Don't drop a single one! Good, 'AI' is coming soon. Now stand on one leg. Keep balancing. You're looking a little wobbly, keep balancing. The iceberg, of course, is climate change. How long can we sustain this juggling act before we drop a ball or two?

We're loath to admit it, but the world is not set up for ape-brained meatsacks any more. There are too many big risks to juggle, and our ancient evolutionary design renders it impossible for us to get our act together and be sound guardians of the planet, or safeguard the future of intelligent life for many centuries to come. Unless we upgrade our cognitive functions and embrace the transition to a more-than-human state, there's a good chance we will exit this blue marbled stage watching cat videos while the world burns.

Readers who are familiar with transhumanist ideas will note that there are lots of interesting topics and technologies I don't explore in detail in this book. Artificial superintelligence is a big one. Other subjects I'm relatively quiet about are molecular nanotechnology, the prospect of hive minds emerging, whole brain emulation (or mind uploading), singletons, the great filter, paradise engineering and wireheading. If you're curious to learn more about transhumanist and longtermist ideas, I'd encourage you to start exploring some of these terms as a next step.

These are fascinating and important topics. The reason they're not heavily featured in this book is because I'd broadly categorise them as event horizon subjects. A world that has unleashed these technologies at scale would be radically different from the present – in many cases, unfathomably different, as the subjective nature of experience for the dominant kinds of intelligent beings would be beyond our present ability to comprehend. These topics are part of the 'what next?' that I hope many of you will be thinking about at this book's conclusion.

My aim here is not to tackle possible futures across short, medium *and* long timescales. I think what most people care about right now is making sense of a world that seems ever more complex, technologically infused, and confusing. Talking about casting off our vulnerable meatsacks (some people find this term for 'body' cringey, but I like it) and uploading our minds to the cloud is too far out for most people – especially those who are thinking about transhumanism for the first time.

It would also be unwieldy to try to unify a discussion about the ever-deeper melding of human bodies and minds with technology in our lifetimes, with speculations about what superintelligent beings might be getting up to in a thousand years' time. Instead of leaping ahead to those posthuman possibilities, we'll remain grounded in this transhuman era, because it is in these transition times that we are living – and right now we are responsible for making big collective decisions that could determine whether any of those more speculative futures emerge. Best to get those decisions right first before we worry too much about how to colonise the stars, or save copies of our minds in the cloud.

The human story

This brings us to the human story. As unstable as our present world and lifeways are, we still have to focus on those core human pursuits like earning a living, building communities, and fulfilling our social, emotional and sexual needs. But we're being drawn ever faster into digital and virtual lives that don't deeply fulfil our hardwired drives and have seduced us further away from the in-person social realms where many of these needs were once fulfilled more readily, albeit imperfectly.

Of course, we're not the first humans to have found ourselves living in a world that feels like it's spinning out of control. In 'Stanzas from the Grande Chartreuse', the British poet Matthew Arnold described living through the transitions of the 19th century as feeling like:

Wandering between two worlds, one dead,
The other powerless to be born,
With nowhere yet to rest my head.

I think many of us can relate to this feeling today as we journey from a human world towards a posthuman future. The chasm between our evolved biology and our modern lifeways is widening faster than ever, and at a rate that will not be sustainable for much longer. Throughout this book I will refer to this chasm, and discuss its related social and psychological effects, under the banner of 'the two worlds problem'. You can think of it a bit like a game of tug-of-war. As the ape-world pulls us by one arm and the cyborg world tugs on the other, we're feeling the pull at our core more intensely than previous generations. It's a tussle that's already broken some of us and has put cracks in a whole lot more.

The human story is about what it's like to live through this push-pull period, and how many of the steadfast pillars of what we think of as a human life-script are set to change. What does the future hold for sex, death, marriage, monogamy, work, procreation, gender, and our basic expectations of what it means to be human and live a good life? What's it like to be a young person today, and is it still possible for a majority of us to pursue the 20th-century dream of a home, partner, stable job and kids?

This is a book of stories within stories. It dips into some of the topical issues and phenomena of the day, which are often explored in isolation: in books about the end of work, the new age of personalised medicine, the rise of virtual reality, the youth mental health crisis, the spike in teens identifying as transgender, and the role that sexbots might play in our future. But these novel threads combine to tell a larger story about a species that's paving the way for a posthuman future. They're rarely seen as the stepping stones that they are, leading to the kind of future that's so far removed from what we know that we can barely imagine it.

Where's it all going?

None of us made a conscious decision to live amid a rapid and major evolutionary transition, a moment when the cultural, technological and economic machinery of the world is hastily expanding into undiscovered futures of greater promise and peril. In our own way, we're a bit like the children of the industrial revolution working in mines and factories, their lives turned upside down, their cities shrouded in smog.

The industrial revolution was good for humanity overall, but terrible for many of the people who lived through it, and terrible for millions of humans in the aftermath too. The same is true of the information age and the transhuman era. But that doesn't mean we can stop the transition from unfolding, or that if we could, we should.

My challenge is to convince you that in the end, and probably sooner than you think, the ape world will lose its grip on us, and the cyborg world will prevail. Far from being the worst outcome, there are many versions of such a future that could be far better than the present and more rich and surprising than we can readily imagine. I also hope to convince you that if you care about sustainability, and want to help build a brighter future for generations to come, you need to start thinking bigger than humanity, bigger than biology and bigger than this planet.

As confronting as some of the ideas in this book are, I hope they spark new questions and get you thinking about how some of the disruptive changes we're already starting to experience are likely to affect your life, your world and the wellbeing of those you care about. The point is not to get you to believe in a rigid narrative where I portray every detail about the future with pinpoint accuracy. It's the big-picture trends and possibilities that you need to consider – and the unexpected worlds you could find yourself growing up, or growing old, in.

If you think you're going to spend the rest of your life living in a human world that looks much the same as it does today, and magically remains stable and sustainable for centuries to come, you are likely gravely mistaken. There is only one way to seed a dynamically sustainable future for intelligent life. We must cast off the shackles of our ape-brains, upgrade, and give rise to the superhuman beings of the future. And time is of the essence.

PART I

HOW TO THINK LIKE A TRANSHUMANIST

1

PREPARE FOR
FUTURE SHOCK

*Today the world changes so quickly that in growing
up we take leave not just of youth but of the world we
were young in.*

Peter Medawar, *The Strange Case of the Spotted Mice*

You blink and find yourself in London in 2022 – a thriving,
modern metropolis. Your head spins as you process the
looming skyscrapers, the gentle roar of the Underground,
planes flying overhead, and neon signs and billboards
emitting dizzying flashes of colour and light. Horns honk as
cyclists weave through lanes of heavy traffic. If accidentally
stumbling into the path of a busy motorist doesn't kill you,
the shock of arriving here might. That is, if you were born
in 1750, before the industrial revolution kicked off and
transformed human civilisation irrevocably.

Waking up here is frightening. Have you died and
gone to hell, or been possessed by demons? As the writer
and illustrator Tim Urban points out, absorbing 270 years
of change might be enough to give you a heart attack, or
trigger a psychotic break. But a person from 1500 wouldn't
experience anything like that degree of shock if they woke

up in 1750. Leaving aside the weirdness of having travelled through time, there would be no rupture in the fabric of their reality, as not that much had changed.

Sure, the world map would look a little different, and the Church of England would have to be explained. As Urban notes, our traveller from 1500 would surely 'be impressed with how committed Europe turned out to be with that new imperialism fad' and 'would learn some mind-bending shit about space and physics'.[1] But he could probably assimilate and go about his everyday life just fine.

Now step out of your time-travelling shoes and imagine interacting with a visitor from our past. How would you explain a smartphone to someone from 1750? They'd be confused by everything! Where did all these people come from? How can those metal birds be flying? You want to inject me with a *vacc* ... what now?[2] They'd be amazed to see so many women covered in jewels. Even their clothes and shoes are embedded with shiny, colourful treasures. As a child skips by in pink sequined sandals, our visitor marvels that these people must know some *astonishing* secrets of mining – and he's right, because plastic is a derivative of fossil fuels.

Convincing him to take Big Bang cosmology, germ theory, or evolution by natural selection seriously would be near impossible – it's hard enough to convince everybody in the 21st century. And it's probably kinder not to mention nuclear weapons and global warming. Or show him footage (wait, there are moving pictures?) of robots breakdancing, or SpaceX landing a Falcon rocket.

Every facet of modern life would seem unbelievable. There are magic thrones that make your poo disappear at the push of a button. Loaves of perfectly uniform bread line indoor 'super' market shelves. They're always there, every day, and you can buy them for next to nothing.

Many so-called poorer people are as rotund as princes and kings. They don't look very poor to our friend – after all, they have their own jewels, horseless carriages and even magic thrones. The old people really are something too. How are so many of them living this long?[3] It must be all the strange foods they eat from those rustling packages. And every kind of delicacy you can imagine is always in season – all the colours of the rainbow!

If he made it this far without losing his mind, our visitor would likely still conclude: *this can't be real*. 'At the next intersection, continue straight', says a disembodied voice from a glowing rectangle. At last, some guidance – but now he knows it's a hallucination. God certainly isn't a woman.

The Great Acceleration

Our friend has been on a heart-stopping journey through a very unusual slice of time. Over the past two and a half centuries, so many mainstays of human lifeways and societies have been radically transformed. There is no clearer way to convey this than visually.

Most of human history looks boring if you graph any key metric of progress. You end up with something like a flat horizontal line. But from the industrial revolution onwards, especially since 1950, blink – *and bam*. You're in a new reality that emerged practically overnight. The Australian climate scientist Will Steffen calls this sudden post-industrial uptick in growth, consumption and environmental transformation the Great Acceleration.[4] Acceleration is a defining feature

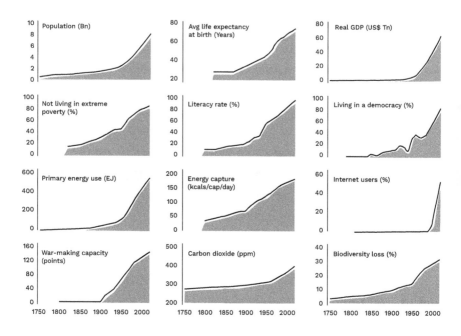

Global trends since the industrial revolution.
Data from multiple sources.[5]

of what many scientists and academics are now terming the Anthropocene – a new epoch in the history of the biosphere, where humans are the dominant force.

I wish I'd seen graphs like these in high school. Back then I hated maths. Other than probability, which I could immediately see was useful for understanding how things are likely to play out in the real world (hence I've never bought a lottery ticket), nobody could ever explain to me what the rest of it was *for*. Hours were spent writing out sums and algorithms in neat rows and solving problems. But as soon as the bell rang, we slammed our books shut and immediately detached from this abstract numberverse – minds drifting back to the things that felt like they mattered in life. Finding the value of x didn't seem like one of them.

It took a while to click, but I got there eventually. *Numbers tell stories too.* And the Great Acceleration trends help frame one of the most important stories of all time. Our species has been rapidly acquiring superpowers in the form of unprecedented levels of control over our bodies, brains, built environments and ecosystems. Overall, the world is richer than it's ever been, rates of extreme poverty are at historic lows, and global child mortality has more than halved in the thirty years since I was born.[6] We're more urban, globalised, literate and interconnected; we produce and consume more than ever; and we're information-rich in a way that's totally revolutionary, but hard to quantify through metrics like GDP and surveys about happiness.

Of course, promise never comes without peril and not all modern trends are positive. Humanity's capacity to inflict damage in war has escalated dramatically, as we've invented new and more powerful technologies that can be weaponised – from machine guns to nuclear weapons, lethal autonomous weapons and their parent technology, artificial intelligence. We're also hounding other species to extinction at an exponential rate, and atmospheric greenhouse gas concentrations are soaring – I've only charted CO_2, but methane and nitrous oxide follow the same trend.

So, why all these sudden spikes at the exact same time?

How we got to now

The industrial revolution created a fundamentally new state of affairs for humanity. It took hundreds of millions of years for photosynthesising plants to accumulate in the earth faster than they could decay and turn into fossil fuels. But humans released that mega-dose of energy practically overnight, harnessing it to power super-large, super-dense, super-educated and super-powerful civilisations. Though we often focus on the social changes, they all piggybacked on top of the energy bonanza we unlocked by harnessing fossil fuels – first coal, then increasingly oil and natural gas.

Why open a book about the future with a history lesson? For the simple reason that we can't understand the present, let alone think about what comes next, without knowing how we got here. With each leap forward in energy capture (think horsepower, and more caloric energy from crops during the agricultural revolution), human civilisations have grown larger and denser. They've also relied on more complex divisions of labour and systems of governance. More people results in more brain power and a burgeoning labour force, which enables us to do impressive new things at an accelerating rate – like build cities, roads and global trade networks.

Once humans began using fossil fuels as a major power source, we could support larger populations without running into the resource limits that forced earlier civilisations to stagnate or collapse. Finding new ways to harness more energy from the environment gave us the power to expand and innovate at a new pace. Breaking free from the historical limits constraining our numbers, population growth sky-rocketed, technological evolution went gangbusters, and a new reality was born.

This modern energy revolution also unlocked the next major paradigm of change: the *information* revolution.

Energy powers brains. Human brains figure out how to harness more energy from their environment, and populations increase. With more thinking power, civilisations invent new information storage methods (from the oral tradition, to writing, print, mass media and the world wide web), and generate and exchange ideas at an accelerating rate.

What globalisation has done for humanity, connecting us through print, trade, migration and travel, *digitisation* is doing faster. It's integrating mind with mind in a deeper fashion – through high-resolution photo and video capture, near real-time language translation, and the instant sharing of content across the globe. We've seeded an extra-human layer of thought, storing ideas, knowledge and experiences outside the human brain. In its modern incarnation, this layer is sometimes referred to as a digital second skin – and we're infusing it with ever more intelligence and power. The next step is to bring it to life.

Nobody knows how far humanity will progress on this superhuman trajectory. But the proliferation of intelligence on our planet – from small bands of hunter-gatherers to a dense, global civilisation, leading to the rise of computers and networked artificial intelligence, is set to define the next chapters of the human story – and at a pace that might take us by surprise.

The second half of the chessboard

Most authors who attempt to explain exponential growth trends in evolution and technology borrow from the American engineer and inventor Ray Kurzweil. It would be hard to better some of his explanations, so why reinvent the wheel? What follows is a very 'gisty' recap of the concepts that are most relevant to the arguments in this book. I recommend Kurzweil's book *The Singularity Is Near* for further reading if you're interested. Or for something simpler, check out Tim Urban's pair of blog posts on Wait But Why website discussing 'The Artificial Intelligence Revolution'.[7]

I vaguely remember learning about exponential growth in school – which you can see in most of the Great Acceleration trends. It's what you get when you double your base number at a regular interval, rather than increasing by a fixed amount, step by linear step. The compounding growth is why change suddenly starts to happen very fast, in defiance of our intuitions. But looking at curves without context on pages of graph paper as a teenager, it seemed like just another boring, abstract pattern.

It was only after reading the works of biologists, physicists, historians, engineers and economists that I learned that being able to think in terms of exponential change, or recognise it in a dataset, can help you make sense of so many important things – from the expansion and collapse of past societies, to our prospects for becoming superhuman, and why you need to mobilise very fast when a novel pandemic emerges, with infection numbers doubling every handful of days.

You might have heard a version of this fable before. It's the one about the Chinese emperor and the chess board. According to the tale, a trusted counsellor presents the Emperor of China with the splendid new game of chess (a game that was invented in India, where some versions of this tale are set). The emperor is pleased and asks his subject to name his reward. The counsellor asks for something very simple: one grain of rice for the first square of the board, two grains for the second, four for the third, and eight for the fourth, doubling each square until the board is full. Sounds reasonable – a few handfuls of rice for his efforts.

Only that's not what happens. First the emperor doles out a sprinkling of rice, then a handful, then a bucket. But by the time he's halfway through the board, the mood in the room has darkened. He's been tricked! The emperor owes the counsellor two billion grains of rice and counting. They never make it to the end of the board, as sixty-three doublings would put him in debt to the sum of 18 million trillion grains of rice.

Many writers, from Ray Kurzweil, to Matthew Yglesias, and Erik Brynjolfsson and Andrew McAfee, have used the chessboard story to make the point that *much* more change happens in the second half of an exponential sequence than the first (in fact, fully half the change happens in the final doubling from square sixty-three to sixty-four). I think it's helpful to pause and reflect carefully on this point, because even when we think, 'yes, that makes sense', we don't really internalise it. Our brains weren't designed to. By the second half of the chessboard, you're hitting scales where the human mind breaks down. We're talking more rice than it's possible to grow on Earth, even if the entire planet were covered in arable land.

So, what if humanity has recently stepped into the second half of the chessboard, in terms of our cultural and technological development? This prospect sounds more plausible when we contextualise the present in light of what's come before. Humans spent 200 000 years as hunter-gatherers, 10 000 years as agriculturalists (5000 of those living in civilisations with cities, states and empires), 250 years in the industrial era, and a few decades in the information age. The pattern's not hard to spot. Major shifts in human lifeways and forms of social organisation are happening in ever more compressed timeframes.

But nobody expects a fundamental upheaval of human lifeways in *their* lifetime. That's because we're a species of linear thinkers who try to make stories fit neatly into human-friendly arcs. When you watch a movie, drama and tension build gradually, and characters work incrementally to address the conflicts in their lives and worlds. Who would ever go and see a film where, for the first ninety minutes, nothing happens? Then in the last thirty minutes a character appears, sits, stares and maybe pulls up a chair. Then in the final minute all the drama, complication, action and resolution are crammed in so fast that you can't parse or absorb any of it.

That's not how stories are supposed to unfold. Stories are supposed to be linear – one damn thing after the other. So if our minds find themselves struggling to process unfathomably rapid change, we assume it must be the story that's deficient, not our brains. As linear thinkers, we tend to ignore an exponential trend in the early stages and discount its ability to overturn our normative reality in the blink of an eye – as so many leaders did in the early (and not-so-early) days of the Covid-19 pandemic, and so many investors did in the early days of the internet, before the dotcom boom. Instead of planning for rapidly compounding disruption, we look to the status quo for comfort and continue to imagine that the degree of change we experienced yesterday will hold steady tomorrow. Time and again we're blindsided as a new reality dawns.

So now it's time to entertain the possibility that we're the ones about to be catapulted into the future from the 2020s. Only this time, no time travel is required. There are no guarantees, but if progress continues, we could live through changes as mind-bending as those our time-travelling friend experienced, or even wilder, in the next few decades. Another compression, another rupture in the fabric of reality.[8] And like him, there's a good chance we won't see it coming.

Pocket-sized supercomputers

One of the most famous exponential trends of all time is known as Moore's Law. It's important because it helps explain why computers keep getting smaller, cheaper and faster – which in turn has allowed us to carry affordable smartphones, video call friends and family around the world, outsource navigation to machines (and even let them take over for a bit when we're driving) and ditch our alarm clocks, cameras, calculators, calendars, dictionaries, and other now-extraneous physical stuff. This trend of digitisation and dematerialisation is part of why people sometimes say that 'data is eating the world'.

Moore's Law loosely refers to the half-century-long norm where computing power doubles roughly every two years for the same cost. That has been due, in large part, to being able to cram more transistors on to a silicon chip – though today we're squeezing out more efficient computation from parallel processing, better software and cloud computing.[9] It's also worth noting that there are four previous growth paradigms in computing, which have built on each other in an exponential price–performance trend since 1900.[10]

Here's a reminder of what Moore's Law has made possible in recent history. If you were in the market for a top of the range computer in the 1960s, you'd need a spare couple of million dollars. A garage-sized room to park it in would be essential too, which only the handful of government institutions and universities that bought them were able to wrangle. What did they get in return for their investment? Number-crunching machines with a few megabytes of memory and about 100 000 times *less* processing power than your smartphone. These colossal beasts didn't even have as much memory as a modern pocket calculator.

What regular person was ever going to want, need or be able to afford a computer? Ken Olsen of the Digital Equipment Corporation (DEC) famously proclaimed in 1977, 'there is no reason anyone would want a computer in their home'. This was the same year Apple released the Apple II personal home computer, which sold over two million units over the next seven years.[11] In the decades that followed, we've gone from scepticism that there would be a market for personal computers to a first-world norm of owning multiple devices and living in environments of ubiquitous computers, sensors and digital connectivity. The unfathomable became omnipresent overnight.

Now consider Warren Buffett, who at the time of writing is the world's eighth richest person. In 2012, with a net worth of $46 billion, he famously said of the internet, 'I would gladly pay half my net worth just to have that kind of information available to me'.[12] Yet neither he, nor anyone else, needs to fork out tens of billions of dollars to have this repository of information at their fingertips. While the gains of technology and industrialisation have not been evenly distributed, exponential technologies become cheap and accessible fast, which speeds up democratised access, turning an ever-larger proportion of the global population into 'information billionaires' by the standards of even the very recent past.

Sure, the prolific spread of information hasn't solved all the inequalities in our world. But it's still remarkable that the internet and modern computers enable widespread access to novel riches that most of us take for granted. Once upon a time, we used to buy rolls of film and thread them inside our cameras. We'd then pay to have photos printed out and store them in albums. Having a million photos printed out at 25c per print would cost you $250 000. Today you can easily store a million photos on a 1 terabyte hard drive, which you can buy for $100.

Moore's Law has been a crucial trend for humanity, because it's enabled so many game-changing technologies to emerge and become affordable in a blink. Technology journalist David Rotman at *MIT Technology Review* notes that, every year, the publication singles out the ten most important breakthrough technologies of the past twelve months. Almost without exception they 'are possible only because of the advances described by Moore's Law'. In 2020, featured breakthroughs and technologies that piggybacked off this trend included: AI-infused smartphones and watches; improved climate monitoring and modelling systems; small, cheap satellites; quantum computing advances; and the development of new anti-ageing therapies and personalised drugs.[13]

In mere decades, computers have become smaller, cheaper and more powerful at a rate that defies our intuitive comprehension. For the most part, we haven't got our heads around where this trend could take us. As the scientist and engineer James Lovelock pointed out in his 2019 book *Novacene*, 'if you think doubling every second year is not that fast, then think again, because it means a thousand-fold increase in twenty years and a trillion-fold growth in a lifetime of eighty years'.[14] We can just about wrap our heads around a computer that's a thousand times more powerful emerging in twenty years, but computers that are a trillion times faster and smarter? I can't picture a trillion of *anything*: stars, apples, grains of sand. The number's just too big.

But let's take a breath. I'm not trying to convince you that Moore's Law – or a larger sequence of historical exponential growth trends, of which it's one – constitutes a future prophecy on which all our predictions should rest. Nor am I saying that if we just extrapolate out from some exponential curves we'll get indefinite progress, with superintelligent machines colonising the stars until they use up all the resources in the universe (assuming they don't find ways to circumvent the known laws of physics). That's possible, but so is an all-out nuclear war, or a deadly pandemic that drives humanity to the brink of extinction.

The farther ahead we project, the more likely it is that there'll be some surprising twists in the tale. Anyone who says they've got a great model and some awesome graphs that neatly predict the future of anything – whether it's the price of Bitcoin next year, or when the housing market will crash by 40 per cent – is speculating. Models are worth making, and speculation is an important activity. We can use these tools to update our confidence about what *might* happen. But there are always the known unknowns and the unknown unknowns that can knock a neat model, encoded with contemporary human assumptions, off course.

The reason the present is so remarkable is because rewriting the fundamental constraints of human nature – ageing, embodiment, and limited bandwidth and brain power – looks poised to become possible for the first time. I'm not claiming that computers will *definitely* match or exceed human intelligence by a specific date (or ever). But absent any major setbacks this century, I'd bet that artificial intelligence will become smarter than us and emerge as the most powerful intellectual force on the planet. Most experts in the AI field think it's more likely than not that artificial intelligence will reach and exceed human capabilities this century.[15]

The essayist and science fiction luminary Arthur C Clarke saw this coming half a century ago, writing in *Profiles of the Future* in 1962:

> This is one of the greatest – and perhaps one of the last – breakthroughs in the history of human thought, like the discovery that the Earth moves round the Sun, or that Man is part of the animal kingdom, or that $E = mc_2$. All these ideas took time to sink in, and were frantically denied when first put forward. In the same way it will take only a little while for men to realize that machines can not only think, but may one day think them off the face of the Earth.

Decades later, the physicist Stephen Hawking echoed Clarke's sentiment, declaring in 2014:

Once humans develop artificial intelligence it would take off on its own and redesign itself at an ever increasing rate. Humans, who are limited by slow biological evolution, couldn't compete and would be superseded.[16]

Coming, ready or not

It took anatomically modern *Homo sapiens* 200 000 years to invent computers. It will not take today's computers and software 200 000 years to invent their own superhuman successors. The deep learning AI AlphaGo Zero only needed three days to teach itself the game of Go from scratch, with no knowledge of the rules, and defeat the AI that beat the champion player Lee Sedol in 2016. After forty days, AlphaGo Zero had beaten every Go-playing AI system and was unbeatable by human or machine. From blank slate to superhuman in days, and super-superhuman soon after. That's the pace at which networked technology can evolve.

Unless progress in modern societies and information technologies slows substantially, or grinds to a halt, the next step over the coming decades won't be a better smartphone with a better camera, it will be algorithms and machines that increasingly exhibit human or superhuman levels of intelligence in at least some domains – that are masters of natural language processing, that can truly understand us, and perhaps eventually be the size of a blood cell, live inside us, and monitor and repair damage in our bodies in real time.

The end game of modern technological innovation is not to build champion game-playing algorithms or superlative driving machines. It's to unlock the mysteries of cognition and consciousness; solve ageing and involuntary death; build sentient, intelligent machines (which, if we're lucky, we can bootstrap ourselves to and become radically smarter, happier and less destructive); and spread intelligence to other parts of the universe.

It's often said that humans tend to overestimate how much a technology will change the world in the short term (I'm still waiting to have an autonomous car drive me home while I take a nap), while massively underestimating how profoundly it will change things in the long run. I think this sentiment is correct and that artificial intelligence will become an archetypal example of the point in our lifetimes.

Technologies with the potential to radically upgrade and transform our species *already* influence how we live and what we value, like the algorithms that determine what we read and see on social media, match us with potential partners, 'tune' our faces in photos to help us conform to modern beauty standards, and in some cases assist modern medical practitioners to make sound diagnoses and personalise treatment.

Soon, these technologies will influence every facet of our lives, extending them far beyond the known upper limits of a natural human lifespan, eliminating many diseases, propelling us into a post-work economy, exponentially accelerating the pace of research and development, and merging human minds with greater than human levels of artificial intelligence. Ultimately, technology will determine what we, our progeny and our species become in the 21st century and beyond.

Barring any major setbacks, humanity's transhuman transition is inevitable. But it will also involve grave attendant risks that we must do everything in our power to mitigate. This is the most important story of our time.

2

T IS FOR TRANSHUMANISM

Let us blast out of our old forms, our ignorance,
our weakness, and our mortality. The future is ours.

Max More, 'Transhumanism: Towards a futurist philosophy'

I became a transhumanist around the same time that I started growing bored with fiction. The novelist David Foster Wallace once said that 'fiction's about what it is to be a fucking human being', and that was my problem with it.[1] Once you've read enough stories about the universal themes of sex, death, love, loss, birth, discovery, betrayal and pain, there isn't that much more to learn or discover. Sure, there are endless nuances to these experiences and a great writer can still bring unique worlds to life in prose that makes your spine tingle. But you basically know where the story's going. Love conquers all, except when death conquers love. Life's challenging and full of complexity, but it's through our trials and tribulations that we grow. Is that it? People fighting, fucking and dying, and pursuing status, fame and glory, or truth, charity and redemption, in various configurations, set in different lands and times, with nothing new under the sun?

After a while you start to get a pretty clear picture of what it means to be a human being. We emerge because two people, who may have had little intention of producing a tiny person, had sex, and their gametes combined. We become products of our world, with limits and traits circumscribed by our genes. We work, we learn, we love, we lose, we die. I used to really get off on the poignancy of dwelling on that entropic trajectory, as though my soul was communing with a deep truth, which heightened a sense of artistic feeling and communion with a universal core of humanity. It never occurred to me that there could be anything *better*, until I stumbled across a remarkable idea and movement called transhumanism.

What is transhumanism?

If you've heard of transhumanism before, there's a good chance you associate it with TV shows like *Black Mirror*, books like Yuval Noah Harari's *Homo Deus*, films like *Ex Machina*, and people like Elon Musk. These associations are all in the right ballpark, at least if they get you thinking about new technologies changing human lifeways, the aspiration to become superhuman, the rise of increasingly dexterous robots, machines that display human levels of intelligence, and tech entrepreneurs working on projects like colonising Mars and merging human brains with artificial intelligence. All of these ideas and projects are transhumanistic. But without a definition that links them together they don't give a clear enough picture of what transhumanism is, or convey why it's such a game-changing worldview and movement in the 21st century.

At its core, transhumanism is a philosophy and a project of technological transcendence that aims to make us *more* than human. It strikes transhumanists as obvious that humanity could be better. This view tends to rattle the cages of humanists, which is funny because they also think humanity could be better. Humanistic sensibilities have been present in many cultures and time periods, but the term is most commonly associated with Renaissance humanism in Europe, which spanned the 14th to 16th centuries, and the Enlightenment brand of the philosophy that followed. In both cases literacy, learning and self-determination were seen as keys to the progressive development of the human spirit. Enlightenment humanism was more overtly secular though, and profoundly influenced the modern Western culture of individualism, free inquiry, free trade, democracy and scientific meliorism.

The important distinction is that humanists don't aspire to outgrow the human condition; they want to use education and cultural tools to bring out the better angels of our nature. They want us to choose the path of Luke Skywalker over Darth Vader and believe that reason, education, scientific literacy and a cultural emphasis on self-determination, rather than divine predestination, are the best ways to go about that. Transhumanists are on board with all of this, but we think that by using the tools of modern science and technology, we can take the humanist project further and actually transform what it means to be human, tackling those final frontiers of human biology and cognition to solve problems of happiness, sustainability and prosperity on a much grander scale.

On this broadest level our goals overlap with those of most of our fellow humans. Not many people think we should be striving to *degrade* humanity, celebrate disease or death, or stop caring about any of the big problems and challenges that plague us. Most people want the world and human lifeways to improve. Where transhumanists differ is in the methods we advocate (like using advanced information technologies to create synthetic life, edit human DNA, 3D print replacement body parts and build increasingly capable forms of artificial intelligence), the degree of change we aspire to (including radical life extension, intelligence augmentation and mind uploading), and the limits of what we consider to be possible.

What are the limits? To the best of modern knowledge our fundamental constraints appear to be the laws of physics. Anything that is in keeping with the known laws is theoretically possible – like creating molecular nano-assemblers that allow you to produce just about anything at a nominal cost (food, insulin, fuel, diamonds) using atoms as raw materials. Or reprogramming a human body to age at a slower rate, or to revitalise and become young again. When it comes to any transformative project that could enhance our lifeways, help us to understand ourselves and the world better, or mitigate sustainability risks, transhumanists think it's worth giving it a crack. Even if we're no longer the same beings we started out as at the end of it.

Transhumanism is hiding in plain sight

Elon Musk is a transhumanist, though he doesn't wear the label. People often assume that Musk's great passion project is to fly to Mars in a rocket, set up a fledgling human colony and die there. If true, I think that's a pretty dumb idea. He cops some flak for diverting money and resources away from human sustainability challenges on Earth and pursuing travel to Mars, as if it's all about building some kind of high-status travel experience for the rich as a self-indulgent lark. I don't think that's what he's doing at all, and the biggest clue lies in one of his other projects: Neuralink.

Neuralink is a brain–machine interface (BMI) company. That means they're working on ways to connect human minds with computers. First-stage applications are medical and relatively non-threatening. BMIs are already used to treat patients with Parkinson's disease and other neurological conditions, alleviating symptoms like body tremors through deep brain stimulation (DBS). They're invasive – electrodes are literally implanted inside your skull by a surgeon and connected to a battery-operated device that's implanted in the chest – so they're not exactly mainstream or sought after as enhancement options. But Musk hopes to change that and he hopes to do much more with them than alleviate symptoms of age-related and heritable diseases.

On the *Joe Rogan Experience* podcast in May 2020, Musk spoke candidly about how BMIs could be used to render human-to-human communication completely unspoken and telepathic. Assuming a continued acceleration of progress in the field, he estimated that it might only take five to ten years to achieve such an outcome and roughly twenty-five years to achieve a whole brain interface, which would effectively put a person in the cloud in a state where 'almost all the neurons are connected to the AI extension of yourself'. Musk also envisages the future ability to 'save the state' of your brain, like in a video game.[2] This doesn't sound like the ambition of a man who wants to fly his meatsack to Mars and pop his clogs: game over.

In August 2020, Musk unveiled Neuralink's latest implant, which he described as 'kind of like a Fitbit in your skull with tiny wires'. A small piece of skull is removed and a coin-sized implant sits flush with the rest of the skull. An advanced robot performs the surgery in less than an hour and it can be done without general anaesthesia. The implant has a wireless range of 5–10 metres and connects to your phone via Bluetooth. The device has been successfully implanted in pigs, and the first human clinical trial will recruit patients suffering from severe spinal cord injuries.

The Neuralink team envisages that this device, and its successive iterations, will be able to restore sight, hearing, and the ability to walk in paralysed patients; warn about heart attacks and strokes in advance; and enable humans to control phones and devices by simply thinking. Musk also believes that in the long run, BMIs will play an important role in helping humanity achieve 'some kind of AI symbiosis, where you have an AI extension of yourself, like a tertiary layer above the limbic system and cortex'. He thinks that this symbiosis is important, 'from an existential threat standpoint', as a merger with advanced AI could prevent humanity from being superseded. While there are no guarantees, the values of human and machine intelligence might be better placed to converge if there is a literal meeting of the minds.[3]

So, putting it all together, we've got SpaceX, a company building reusable rockets and accelerating competition in the aerospace industry, driving prices down and catalysing design innovation. Then there's Tesla, which makes cars but is really an AI company working on computer vision, image recognition, machine learning and autonomous decision making, as well as sustainable energy solutions. Then you've got Neuralink, which is focused on the biological side of interfacing with machine intelligence. Musk is clearly very interested in accelerating our transition to a more fluid, cyborgian mode of existence where we seamlessly interface with AI and eventually integrate it more fully into our bodies and minds – he's said as much on many occasions. That's a fundamentally transhumanist aspiration.

So why does Elon *really* want to go to Mars? Don't get me wrong, I think the boyish science-fiction nerd in him truly does want to see it. But he's explicitly said that going to Mars is a hedge against the extinction of terrestrial intelligent life. The way to do that isn't to put vulnerable meatsacks on a planet they're not evolved to survive on and expect them to thrive. It's to send humans there *with* machines, so that seeds of both organic and non-organic life and intelligence can be planted there.

The non-organic part is the most important part though, because it's the most resilient in the long term and will evolve much faster. It's just that we'll need some humans on the ground helping them to get set up. We're much more dexterous than a rover and have better general problem-solving skills than machines for now, hence the need for affordable rockets, and humans to get in them. If we can build environments that are safe for digital intelligence to exist and evolve in, then sending humans to Mars actually does make sense as a way of attempting to safeguard the future evolutionary prospects of intelligent life.

How do you sell that to most humans? Do a Kennedy-esque sales pitch and spruik it as a moonshot that's all about human exploration and the noble conquest of cosmic frontiers. Put humans front and centre as the heroic voyagers and we're much more likely to go for it and get involved in the narrative. Tell us upfront that it's about transcending humanity-as-we-know-it (while preserving the best parts of it, like intelligence) and you've got a much harder sell on your hands.

Big tech is big transhumanism

In our everyday world Elon Musk is an anomaly, but in the tech world he's almost normal. It's part and parcel of being an information-age tech-industry billionaire and thought leader to chase moonshots. A deep understanding of modern information technologies distinguishes people like Musk, Larry Page, Sergey Brin, Mark Zuckerberg, Jeff Bezos, Richard Branson, Bill Gates, Larry Ellison and Peter Thiel from other kinds of billionaires. They also share an innate curiosity about big-picture issues and challenges that affect the entire species and the future of the biosphere and intelligent life. They're a 'type' and they couldn't be more different from the rest of the Forbes rich list: from the Saudi princes, to the Donald Trumps, and the Mars candy empire heirs and heiresses.

I don't think any self-made billionaire is stupid, but information-age tech billionaires possess a rare brand of visionary thinking in abundance. They are some of the few people in our world who understand, to the extent any human can, the complexity of the technological systems that are driving the most rapid and profound transformations in our world. With their tremendous wealth, global influence and technological expertise, these figures are in a real position to have a conscious impact on the development of human societies in their lifetimes. Their involvement with existing tech behemoths – the ones that know us inside out, feed us ads, predict our thoughts and shape our habits – also adds to their unprecedented power to change human societies and the human species. For better or worse.

That wouldn't be so important if they spent their days on massive yachts sipping champagne and indulging in hedonistic pleasures. But these guys are big-picture thinkers who enjoy learning about how the world works, what the biggest challenges of our age are, and how they can play a role in solving them. And many of the projects they're currently spearheading and investing in have a profoundly transhumanist flavour.

Some notable examples include donations made by the entrepreneur and venture capitalist Peter Thiel to the Machine Intelligence Research Institute (MIRI), which focuses on safely developing robust and friendly AI, and the Methuselah Foundation, which promotes anti-ageing research with the goal of 'Making 90 the New 50 by 2030'.[4] Thiel is a longstanding life-extension enthusiast. So much so that his early company Confinity, which later merged with Elon Musk's X.com to become PayPal, was the first in the world to offer cryonic suspension ('freeze your head to save your ass', as transhumanists used to say) as part of its employee benefits package.[5]

Thiel has unambiguously declared, 'I believe if we could enable people to live forever, we should do that'.[6] In the foreword to Sonia Arrison's 2011 book, *100 Plus: How the coming age of longevity will change everything, from careers and relationships, to family and faith*, he wrote, 'Death will eventually be reduced from a mystery to a solvable problem. In this reduction we may hope that human life will achieve a new level of freedom and consciousness'. He continued, 'death was natural in the past, but so was the instinct to fight it. The future only has room for one of them'.[7]

There's also Mark Zuckerberg and his wife Priscilla Chan's pledge to invest 99 per cent of their Facebook shares over the course of their lifetime – valued at $45 billion when the announcement was made in 2015 – in the Chan-Zuckerberg Initiative (CZI). This initiative aims to accelerate research in high-impact areas in a bid to solve some of humanity's greatest challenges. One such project is to map every cell in the human body and attempt 'to help cure, prevent, or manage all diseases' in their children's lifetimes.[8]

In a 2015 town hall Q&A on Facebook, the physicist Stephen Hawking posted the question: 'I would like to know a unified theory of gravity and the other forces. Which of the big questions in science would you like to know the answer to and why?' Mark Zuckerberg replied:

That's a pretty good one!

I'm most interested in questions about people. What will enable us to live forever? How do we cure all diseases? How does the brain work? How does learning work and how we can empower humans to learn a million times more?

I'm also curious about whether there is a fundamental mathematical law underlying human social relationships that governs the balance of who and what we all care about. I bet there is.[9]

Let that sink in for a moment. Zuck wants to map cognition, crack the code of human personality and behaviour, augment learning a millionfold, cure *all* diseases and live forever. Pretty hard to argue he's not a transhumanist. His statements epitomise an aspirational worldview to radically enhance human capabilities, using advanced information technologies. He believes these enhancements are both possible, and desirable. More importantly, he's at the helm of one of the biggest AI companies in the world, which is busily mining some of the largest datasets ever accrued in order to get to know us inside out.

Love him or hate him, Zuckerberg is yet another influential billionaire pursuing a transhumanist future. I say this as a transhumanist, knowing that such an association will likely tarnish the entire idea of transhumanism for many of you who are reading about it for the first time. I say it anyway, because we need to see this phenomenon for what it is, in its full scope and influence, and transparently consider the promise and perils of transhumanist projects in concert.

To be clear, there's no doubt that having any human being – or technocratic oligarchy, however smart – at the helm of such a transition is dangerous. But if somebody's going to exert disproportionate influence in our world, I'd rather it be a very smart person, with a deep knowledge of modern technology, all else being equal. That doesn't mean I worship at the altar of such humans, believe them to be infallible, or think we should abandon our critical faculties and fail to hold them accountable if and when they misbehave.

But this isn't a book about the legitimacy of big tech. The point of this section is to highlight how big-tech projects are propelling us ever-faster towards a posthuman future, and it would be a glaring omission not to mention Alphabet/ Google. Co-founders Larry Page and Sergey Brin were the first to create a moonshot division in the big tech world – called X (formerly Google X). Projects that started in this division, have been explored there, or remain there today include Waymo self-driving cars; the company's machine learning division, Google Brain; Verily's smart contact lenses that aimed to measure diabetics' glucose levels using their tears; the California Life Company (Calico), which is accelerating fundamental research in human biology and ageing; and Project Loon, which has sent fleets of balloons into the stratosphere to connect remote and rural communities to the world wide web, with the ultimate aim of integrating the last few billion humans who do not have internet access into the global digital brain.

Google also acquired the robotics company Boston Dynamics in 2013, which has famously developed highly mobile and dexterous robots that can walk, run, jump, dance and even do backflips – though they sold the company to SoftBank in 2017 (who, perhaps tellingly, sold it on to Hyundai at the close of 2020). While commercial demand for such bots remains limited, they have challenged assumptions that it would take much longer for robots to master basic physical movements that humans find easy. Bots of this ilk hint that it might be possible in the coming years, or decades, to unravel Moravec's paradox, which states that 'it is comparatively easy to make computers exhibit adult-level performance in solving problems on intelligence tests or playing checkers, and difficult or impossible to give them the skills of a one-year-old when it comes to perception and mobility'.[10]

Google also outbid Facebook to acquire the AI company DeepMind in 2014 for approximately $650 million.[11] DeepMind is famous for creating AIs that can defeat the best human Go players, and for quickly developing next-generation deep-learning AIs that taught themselves to play the game from scratch with no prior knowledge or training and rapidly defeated their own superhuman machine predecessors – effectively evolving from neonate to super-superhuman in mere days. Of course, the real goal of DeepMind, as stated on the company's website, is 'to push the boundaries of AI, developing programs that can learn to solve any complex problem without needing to be taught how'.[12] In short, they are trying to build an artificial general intelligence (AGI), which means a kind of AI that could do just about any task as well as, or better than, the average human and switch seamlessly from task to task. DeepMind is busy inventing a future where anything you can do, a bot can do better.

Google even famously hired one of the world's most prominent transhumanist thinkers, Ray Kurzweil, as its Director of Engineering in 2012. Kurzweil is an inventor, engineer and polymath who has received the US National Medal of Technology and been inducted into the National Inventors Hall of Fame. He popularised the concept of the technological singularity (the idea that we're heading for a rapid intelligence explosion due to exponential improvements in information technology) and has been at the forefront of life-extension experiments, famously taking dozens of carefully chosen supplements and medications each day to slow and circumvent the processes that contribute to ageing. Journalist Dawn Chan rightly highlighted the significance of Google's decision to hire him when she wrote in the *New Yorker* in 2016, 'pause to consider the fact that the second-largest company in the world (by market capitalization) has a director of engineering who believes that humanity will conquer death'.[13]

We're apt to forget how novel this breed of thinker and social influencer is. Back in 1918, the richest people in the world were monarchs, aristocratic and dictatorial rulers, and entrepreneurs who made their fortunes in first industrial revolution industries like oil, railroads and mining. In 1918, America had 18 billionaires (in 2017 dollars).[14] A century later, in 2018, America had 585 billionaires – the highest number in the nation's history to date. The richest American in 1918, the oil magnate John D Rockefeller, had a net worth equivalent to around $21 billion today, but the second richest American, Henry Clay Frick, had a net worth that paled at only $3.9 billion.[15]

Depending on which day you check, today's richest American, Elon Musk, who is also the richest person in the world, has a net worth of $241 billion.[16] Between them, America's top ten information-age tech billionaires have a combined net worth of over half a trillion dollars (or 2.5 per cent of the US economy and over half a per cent of the global economy).[17] This matters. For the first time in history, a large number of the world's most powerful elites are simultaneously highly intelligent, highly educated, extremely tech-savvy and sympathetic to transhumanist ideas and agendas.

With such capital and power behind it, I believe transhumanism will be to the 21st century what capitalism and socialism were to the 20th: one of the most important and influential ideologies of our time. We need to name it in order to recognise and grapple with its importance and influence in our world, not just today, but tomorrow, and decades and centuries hence.

The 'problematic' stereotype

The trouble is, transhumanism has an image problem. If you had to design a stereotype with the express purpose of making it difficult for the average person to identify with and relate to, it would be hard to improve on the readymade that is 'the transhumanist'. Picture a white, male, libertarian-leaning fellow with an above-average IQ who reads prolifically about all manner of odd topics, like space travel, the probability of life existing elsewhere in the universe, how to build an artificial brain, gene editing and cryonic suspension (being frozen after death in the hope of being re-animated in the future).

This person likes big-picture thinking and philosophy and thinks deeply about complex ideas and global problems. You'll most likely find them in a university or a tech company. They tend to care a lot less than the average person about emotive topical issues, like which politician or celebrity said the darndest thing today, and they love thinking about what could be, rather than what is. They might not be a strict utilitarian (more on that shortly), but they'll know a lot about utilitarian philosophy and place a high premium on logic and rationality.

Most people are more captivated by everyday minutiae, and more governed by emotion when making decisions. So it's not surprising that when they hear about speccy tech guys geeking out on weird futuristic ideas, especially the kind that sound potentially threatening to the warm, gooey, human world we know and love, their gut reaction is aversion and their first instinct is to find ways of deriding the archetype (too male, too nerdy, too 'aspie') so that any ideas a person of this ilk conveys can be summarily dismissed.

I confronted this reaction when I explained trans-humanism to a meeting room full of academic historians in 2019, watching their brows furrow in unison as I spoke. By way of background, I noted that the first groups of organised transhumanists in the 1990s were young, a bit kooky, fond of neologisms, predominantly white and male, and often adopted new and futuristic names like Max More, Natasha Vita-More, and TO Morrow (or Tom Morrow, get it?) to symbolise their commitment to their forward-looking ideals. But I was careful to emphasise that, overall, they were also an extremely clever bunch – many had PhDs from top-tier universities, and in quite a few cases they were, or went on to become, world-leading researchers, scientists, entrepreneurs and inventors.

Among them was the co-inventor of public key cryptography Ralph Merkle, the artificial intelligence pioneer Marvin Minsky, the father of nanotechnology Eric Drexler, the Carnegie Mellon roboticist Hans Moravec (from whom we get Moravec's paradox), the entrepreneur and inventor Ray Kurzweil, and the philosopher Nick Bostrom, who has probably done more than any other thinker to bring transhumanist ideas into mainstream consciousness in recent years through his bestselling book *Superintelligence*.

The historians were unimpressed. It was as if the minute they heard white, male and into technology, their brains instantly switched off. Different tribe, yuck! One of the more gregarious female senior lecturers chimed in with a wide grin, 'I hate them already!' and the rest of the group chuckled with tribal affirmation.

I tried to explain that it doesn't matter whether you like them personally or not, it matters that their ideas and projects are becoming influential in the modern world and are worthy subjects for historians and the public to be studying and thinking about. Even harder to get across was the point that even if some transhumanist ideas seem misguided, or unpalatable, it doesn't automatically mean that *all* of their ideas can be summarily dismissed as adolescent tech-geek fantasies.

Just like Australian tourists in years gone by, transhumanists are everywhere. You're reading a book by one right now. The trouble for transhumanism's problematic image is that transhumanists like me don't make for juicy clickbait stories about rich male technocrats who want to reduce all aspects of human nature, like love and art, to code. I read a lot about health, nutrition, life-extension drugs and biohacking. But I've never implanted any chips under my skin, haven't signed up to be cryopreserved after my death (though I might do it one day), and don't flinch in horror at the prospect of interacting with other humans in real life.

I'm also not a millionaire, or a billionaire, and I don't aspire to engineer a grandiose tech utopia that turns rich people into gods while trampling the poor into an obsolete heap of atomic dust. Boring, right? Which is precisely why the media doesn't talk about transhumanists like me, or the very sensible and Enlightenment-based ideals of transhumanism. Who's going to read a lengthy academic paper called 'The philosophical roots of transhumanism' when they could eyeball the headlines of articles reporting that: 'Peter Thiel is very, very interested in young people's blood,' and that 'Silicon Valley geeks are "turning their kids into CYBORGS"'.[18]

I've spent years doing academic research on transhumanism (which you can read online for free if you're interested in more details) and I don't know of a single transhumanist who aspires to a future where the haves triumph gleefully over the have-nots. Transhumanists overwhelmingly, if not universally, champion the opposite goal, and often spend years thinking carefully about policies and strategies that could help realise it. It's just that opinions differ about the most effective means of bringing that ideal to fruition, as you would expect in the case of any complex global problem.

The meme humanity needs now

Transhumanism goes by many names and has many faces, and it's often hard to spot. You probably won't hear teenagers in school playgrounds, or work colleagues, saying 'I'm just so obsessed with transhumanism'. You might hear them talking a lot about computers, though. Or robots, or cyborgs, or colonising Mars, or even the prospect of artificial intelligence becoming superintelligent one day, like in the films *The Terminator* and *Ex Machina*. Those memes are replicating and proliferating quite successfully.

We also hear many different calls from different corridors of academia, all in their way suggesting some version of rewriting human nature, overcoming our limitations, and becoming more and better than we currently are. The evolutionary biologist Bret Weinstein is calling for humanity to seize the evolutionary reins and free ourselves from the shackles of a human nature that is not fit to survive and prosper in the modern world.[19] While I suspect Weinstein would not endorse the suggestion that he is a transhumanist at heart, his ideas certainly dovetail more with a transhumanist worldview than any other philosophy or ideology, as far as I can discern.

In molecular biology, we have David Sinclair at Harvard Medical School declaring that human ageing can be slowed and reversed and that we can, and should aspire to, live healthy lives well beyond 120 years.[20] In astrophysics, we have the British Astronomer Royal Martin Rees affirming the inevitability of a posthuman future and declaring that 'organic intelligence has no long-term future' because 'by any definition of "thinking" the amount and intensity that's done by organic human-type brains will be utterly swamped by the cerebrations of AI'.[21] These are also fundamentally transhumanist goals and visions.

A growing number of people in educated circles are now talking about Nick Bostrom's idea that we might be the spawn of a more technologically advanced civilisation and could be living in a simulation. In tech circles, popular culture, and even politics (though the pollies have usually left office by the time they say anything publicly, or else are not democratically elected) we have just about everybody piling on the 'AI is coming' bandwagon now, from Joe Rogan, to Sam Harris, Stephen Fry, Elon Musk, James Lovelock, Barack Obama, Hillary Clinton, Henry Kissinger, Vladimir Putin and Xi Jinping.

This is not a cast of fringe characters; it is a cast of eminent researchers, world-leading thinkers, intelligent social influencers, and leaders of powerful nation states with access to advanced military R&D intel – and there are many more I could have named and quoted, from the late Stephen Hawking, to the physicists Max Tegmark and Michio Kaku, the leading AI researchers Stuart Russell and Demis Hassabis, and Skype's co-founder Jaan Tallinn.

But here's where the transhumanist meme runs into a problem. It's a really broad framework concept that requires us to *simultaneously* think about the big picture and the human story, and weigh up present goals and priorities against future goals and outcomes. Humans don't really like doing this and as we're about to see in the next few chapters, we're not very good at it.

As I mentioned earlier, there's also a natural utilitarian or consequentialist flavour to transhumanism. This means trans-humanists tend to be more comfortable making decisions based on the greatest good for the greatest number. Our biggest concern is whether the overall consequences of a decision are positive, even when there is collateral damage in that equation. This preference conflicts with the blanket principle of 'do no harm' – which, if applied too literally, can cause massive harm. This utilitarian bent may also contribute to many people's aversion to transhumanism, and those who espouse transhumanist ideas.

A 2018 study in the *Journal of Experimental Social Psychology* found that non-consequentialists who refused to break moral codes to ensure a better outcome (by, say, sacrificing one life to save five) 'were consistently viewed as more moral and trustworthy, preferred for a range of social roles, and entrusted with more money in economic exchanges'. Non-consequentialists who valued the wellbeing of acquaintances over the wellbeing of strangers were also preferred as close personal friends and romantic partners.[22] So a majority of people may in fact be primed to think negatively about those who lean in a consequentialist or transhumanist direction.

It may seem odd that transhumanists champion the ongoing goal of *incrementally* improving human lifeways and abilities (we're all for spectacles, antibiotics, vaccines, and tackling malnutrition and poverty), while also advocating more radical changes to bodies and minds in the hope of dramatically increasing human healthspans, augmenting our natural abilities, and one day seeding positive kinds of posthuman beings. But there's nothing odd or contradictory about this.

Everything that contributes to that incremental improvement of human lifeways, healthspans and capabilities is part of the broader transhumanist project. I suspect the reason that most people don't associate lots of those endeavours with the end goal of a posthuman future is because that prospect does not yet feel imminent – more so because those who are pushing our species in transhumanist directions either aren't talking about it overtly most of the time (like the big-tech guys), or don't even know they're doing it (like most members of most governments). As a result, transhumanist phenomena and the idea that we're living in a transhuman era is much less visible and widely understood than it should be.

I think academics, authors and book publishers are also partly responsible for our ignorance of this framework concept. With a few prominent exceptions, they have long shied away from the difficult task of explaining it. I can tell you firsthand how understandable that is, because it turns out it's really hard to do without boring or confusing the heck out of people (and I really hope I haven't done that here). But either way it's worth a crack, because this concept is important.

As a workaround, many authors have coined their own terms to talk about transhumanist phenomena and trajectories, whether it's the Third Industrial Revolution, the Fourth Industrial Revolution (identical concepts; the World Economic Forum simply co-opted the former concept by the economic theorist Jeremy Rifkin and whacked in an extra dividing line), Industry 4.0, the Second Machine Age, the Anthropocene, the Novacene, the Great Acceleration, Homo deus, technosapiens, Homo digitalis, or whatever else. Academics, I implore you: stop coining new terms to stamp your personal brand on widely discussed techno-cultural evolutionary phenomena! The Anthropocene's probably the only neologism that makes sense, because it's a geological term based on empirical changes in the geological record, and follows longstanding naming conventions of the field.

Transhumanism, meanwhile, is not a term or concept somebody made up yesterday. It has a thirty-year history as a coherent philosophy and subject of academic inquiry. The idea of a transhuman era seems like the natural meme to describe the complex set of interrelated transformations (geological, ecological, economic, technological and cultural) that are generating, or signalling, profound acceleration and change in our world. Unlike other terms, it also encompasses the idea of a conscious transformation, fuelled in part by projects instigated by powerful people and institutions, which are *deliberate* attempts to overcome human limitations and render us more-than-human. It's profoundly unhelpful to splinter a coherent concept like transhumanism into a cluster of sub-concepts that the public hasn't learned to associate with an overarching conceptual framework. This creates the illusion that transhumanist ideas and technologies are less powerful and pervasive forces in our world than they really are.

When I hear housemates and friends talking about 'industry 4.0' technologies, which are now being discussed in their workplaces, I despair a little, because this meme is narrow. It always focuses on today's capabilities and today's application of information technologies, or what might be possible in a year or two. It's all about how businesses can use big data and existing algorithms to enhance productivity in narrow domains, and when I talk to people about these technologies they rarely have any thoughts about the long-range adjacent possibilities that could arise from these technologies, or how different industry and human lifeways could be in the coming years and decades. Meanwhile, a bunch of overlapping and related memes that speak of parts of this important story of a transhuman era continue to compete for eyeballs and headspace.

Some remarks on #diversity

Even leading transhumanist thinkers like Nick Bostrom, who composed some of the foundational documents and philosophical outlines of transhumanism, have recently begun to eschew the label – seemingly because of how readily it's associated with white, male nerds and techno-utopian zealots.[23] This is a move in the wrong direction. The reason that many transhumanists have started pretending they're not really *that* any more is because they're scared of being associated with the 'problematic' image we talked about earlier. Yet they remain as avidly engaged with transhumanistic projects as ever, which is another reason why this powerful ideology appears deceptively small and on the fringe.

It's true that roughly three-quarters of transhumanists have historically been men, and this shouldn't shock or disturb us.[24] Half of all people are men, and men are perfectly cromulent people. It's also okay for a group of people to disproportionately share some traits and to not be fully representative of broader society in every way. I wouldn't expect to find many people who dislike cycling among a group of cycling enthusiasts. Nor would I expect to find lots of over-65s protesting for lower university fees, or to hear that a majority of the clients at a nail salon were men. Transhumanists are certainly keen to have more women among their ranks, but I don't think it's terribly difficult to figure out why women tend to be less interested in these ideas *on average*.

Transhumanism is an extremely big-picture worldview that involves thinking about the deep evolutionary history of the universe, life, humans, technology, and the possible futures of the planet and intelligent life. Most people, regardless of sex, don't tend to gravitate to those kinds of ideas or like to spend a long time thinking about them. Why? Because it's not adaptive. Evolution wouldn't select for a majority of humans to want to spend their days thinking about how long it would take to build a Dyson-sphere (or Dyson-swarm), which could harness the bulk of the energy of a star and power a more advanced future civilisation. We'd have all been eaten by tigers long ago if humans weren't pervasively wired to be present-focused, tribal and short-term oriented.

But you especially want the women in your tribe to be present-minded and sensory, because their job, evolutionarily speaking, was to safely bring babies to term, birth, breastfeed and nurture them. Those tasks don't mesh well with having your head in the clouds. Unlike men, women can't make a bunch of babies each year and invest in the ones we feel like investing in, if it suits us. Reproduction for women is slow and very costly, so we're built to approach it more carefully, and invest more consistently in doing it well.

Unfortunately for Palaeolithic women, there was no paternity leave, or childcare centres. There was 'the tribe' and a group of other women to share gathering and child-rearing tasks. It would have been most adaptive to have a group of women who largely enjoyed thinking about day-to-day tasks and bonding with other women over their shared responsibilities. Such women would have had far more reproductive success than women like … well, me. I fear I'd have been far more likely to have been burned at the stake at one time or another, or whatever the Palaeolithic equivalent ritual sacrifice for heresy was. Alas, being put to death for various tribal faux pas isn't terribly conducive to getting your genes into the next generation.

The same goes for men too, of course, with hunting, gathering and warfare. You want most men to be present-focused and not too esoteric. But there's a bit more of a need in the historically male arena for greater variation, including more macrostrategy from time to time – being able to anticipate changes in a niche, or formulating a religion that promoted tribal cohesion, for example. That's a very crude outline of my hypothesis for why most humans aren't drawn to big-picture thinking and transhumanist-type ideas, and why that preference is more commonly expressed in men.

Here's another way of thinking about transhumanist demographics. We're a bunch of outliers, who think a little bit differently from the norm. There's nothing inherently good *or* bad about having traits that make you think a little bit differently. How we judge difference should be based on whether it adds value. The real question we should be asking ourselves isn't: are transhumanists my kind of people who think like me and share all my values? It's does this group of people who might think differently to me have anything useful or interesting to say?

Transhumanists are offering the world something very important at a critical juncture in the human story. We're offering a perspective that doesn't come naturally to most people – one that is more oriented towards the big picture and the long-term future, which humans tend to discount and fail to plan for safely.

Don't worry, we know we're still human and riddled with our share of ape-brained biases. But by tending to be less interested in short-term, local, tribal issues, we're often better placed to contemplate, and work on, the biggest issues facing our species – and not just our favourite pet sustainability issue, like climate change, but thinking about *all* of the biggest challenges and opportunities in concert. This endeavour is in the interests of everybody, even those who don't enjoy contemplating these thorny conundrums, because there will be no short-term concerns to angst over, or invest in, if our planet and civilisations fall apart.

Transhumanists have argued for decades that it is extremely important to look beyond the current wave of attention-grabbing gadgets and contemplate the long-range future of humanity. The earlier we start thinking about the probability of various posthuman or extinction scenarios, and the pathways that could lead us there, the better prepared we will be to intervene and make every possible attempt to try and shape the future for the better.

You don't have to like transhumanists, or want to have us over for a dinner party. But you should be interested in what we have to say, and open to the prospect of having your views about humanity and its future over the next century challenged. If there's one thing our global society can learn from a transhumanist worldview, it's the importance of taking the long view, playing the long game, and minimising existential risks, so that humanity and future forms of intelligent life do not have their potential wiped out. You can take or leave the rest.

3

WHAT'S SO GREAT ABOUT HUMANITY ANYWAY?

There is no graceful aging. All aging is graceless.
There is no dignity to dying. Death is the ultimate
indignity.
Let us stop this self-deception.
In our times the only dignity is in mobilizing intelligently
to overcome aging and death.

FM Esfandiary, *Are You a Transhuman?*

Buttercup: *You mock my pain!*

Man in Black: *Life is pain, Highness. Anyone who says*
differently is selling something.

William Goldman, *The Princess Bride*

My 86-year-old nonna has repeatedly given me the same piece of advice for a good ten years and she always says it in earnest: 'Never grow old!' She has no hesitation admitting that suffering is a prominent feature of her life. She's literally kept alive by a pacemaker and handfuls of pharmaceuticals that she consumes on a daily basis, all of which have side effects she has no choice but to put up with, in addition to her daily aches and pains, limited mobility and declining memory and vigour. She can only shuffle around very slowly, is always breathless, can't chew food properly, and is tired and anaemic. Most of her friends are dead and she's been going to a funeral every other week for years.

She's lucky that she doesn't suffer from dementia or any serious form of cognitive decline. But the downside is that she remains acutely aware of how much capacity she has lost and how little else there is to lose before the game is over. She is terrified, and I am terrified for her. The worst part is that there will almost certainly be more suffering to endure – the little things that don't finish you off but slowly eat away at you until there's no resistance left. There is nothing dignified about this and there is nothing beautiful about it. It is heartbreaking.

Losing loved ones is one of the worst things about being human. And yet we tell ourselves that, as sad as it is, it's also a humbling and important thing because it's part of the circle of life. And *we're* part of the circle of life. And life would be pointless, hedonistic and unfulfilling without mortality and a ticking clock to remind us of its value. The only trouble with that story is that it isn't true. We made it up because we feared the reaper and we didn't have the tools to keep the scythe-wielding villain at bay.

The stories we tell ourselves

Of all the people to have written a consoling myth about human mortality, you wouldn't expect it to be Richard Dawkins. But in *Unweaving the Rainbow* he writes:

We are going to die, and that makes us the lucky ones. Most people are never going to die because they are never going to be born. The potential people who could have been here in my place but who will in fact never see the light of day outnumber the sand grains of Arabia. Certainly those unborn ghosts include greater poets than Keats, scientists greater than Newton. We know this because the set of possible people allowed by our DNA so massively exceeds the set of actual people. In the teeth of these stupefying odds it is you and I, in our ordinariness, that are here.[1]

This is one of my favourite book passages. It is moving and poetic, and it also happens to be true. We *are* lucky in the sense that we have defied staggeringly long odds to get here. Our existence is contingent on trillions upon trillions of historical dominoes lining up exactly as they did, going all the way back to the Big Bang. This is an amazing and humbling thought. But the rousing swell of humility that we feel in the face of the cosmic deep only lingers for about as long as it takes to read the words, 'it is you and I, in our ordinariness, that are here'.

Dawkins' portrait of the human condition is designed to be consoling. It highlights the beautiful view from the mountaintop that science can provide, the sweeping panorama that helps put our existence into context. This view enriches our lives with new knowledge, and rouses our minds with the poetic beauty of what Dawkins terms 'the magic of reality'. As much as I agree that having the chance to glimpse this magical reality is a wonderful thing, we're only focusing on the portrait adorning one side of the coin. What about the many tragedies of reality, like ageing, sickness and involuntary death?

My nonna's youngest grandson, Cassius, told her when he was little that he was going to grow up and become a scientist so he could invent a potion that would stop people from dying. She tells us about the potion a lot, wide-eyed and excited about the idea for a few moments before sighing in resignation that any such invention would come too late for her. That's when she reverts to praying. Would she take the potion if it existed? In a heartbeat, without hesitation.

Like the 17th-century philosopher Blaise Pascal, who is famous for his wager – where he posited that it was smarter to bet on God's existence and be faithful, because spending eternity in hell would be a heck of a punishment if the non-believers turned out to be wrong – my nonna's basically hedging her bets between science and religion. She's also remarkably adept at holding two wildly divergent beliefs about mortality: that it's bad and she doesn't want to die; and that dying is perfectly fine and nothing to fear because she'll go to heaven. She isn't betting on heaven because she thinks it's more likely to be true. Like Pascal, she opts for faith because there's more to gain personally by believing in God if he turns out to be real, and not much to lose by going along with the story if God's a fiction.

I'm one of those irritating people who's more interested in what's true than what's consoling and I've always been fascinated by death. Not because I'm morbid, but because it's part of the framework of life, and to ignore it is to avoid examining the human condition in the round. I've never been capable of hedging my bets when it comes to ageing and death; I've always thought that they're obvious tragedies that we should decry and strive to overcome. Except in cases where dying would alleviate or prevent prolonged suffering – and of course people should be free to choose to die if they wish to.

In addition to challenging the consoling mythology of the circle of life, this chapter tackles one of the most universal sacred cows of the human psyche – a belief that has been remarkably consistent, though periodically questioned, across eras and cultures: that being human, with everything that entails, is intrinsically good and better than the alternatives. As humans, it's not surprising that we're not wired to think deeply about whether humanity is a good thing. If we came to the conclusion that it wasn't so great after all, we might not be as motivated to protect and defend our tribes, stay alive and have more human babies. But we're programmed to do all of those things and we need to keep thoughts that might undermine the success of those endeavours in check.

Unsurprisingly, we're very good at doing this. But if we're really honest with ourselves, we know that humanity has a very dark side and there are lots of things about the human condition we'd do anything to change if we could.

Get down off that pedestal

To be human is to love, or so we like to tell ourselves. There is something special, ineffable and unprogrammable about us. We are more than just the sum of our parts. We are of the earth, of nature, of a great interconnected ecology, and we can think, dream, feel emotions, create and desire. We can hold children in our arms and feel the deep primal bond of unconditional love. We can write poetry, pass knowledge and traditions down through generations, be altruistic, love thy neighbour, and come together to build a better world.

We are very apt to romanticise the best parts of our nature, and like a great PR person or politician, we're very good at downplaying the many ways in which we're just plain terrible. Not to mention hilariously poorly designed. I mean, what's the deal with half of the adults of the species bleeding out of their vaginas for one week in four, for several decades of their lives? Or with most of them at some point painfully squeezing a live human being out of a tiny aperture after carrying it around inside them for nine months? And what kind of engineer would put the frontline tools of the sexual funhouse in alarming proximity to the sewerage system? I'm not the first person to say any of this, but it's worth reiterating for how comically bizarre it all is. We are first and foremost animals who carry many vestigial inheritances in our design that no longer serve us well.

Of course, those are just some of the small indignities and hilarities about being human. Here are some bigger ones. If you really think about it in the broader ecological schema, we are our planet's playground bully thugs. We're nice enough, or behave neutrally, towards species that are allies, or don't have lunch money we want to steal for ourselves. We're even rather good to all those crops and animals we selectively breed and have a symbiotic relationship with – they're much more evolutionarily successful because we select for them and deem their competitors and predators pests and weeds. But when we see a glimmer of currency, in the form of ivory tusks, or the rich payload of caloric energy stored in the flesh of another creature, we pounce and devour them.

We used to do this kind of thing on a fairly modest scale. But in the post-industrial world, where factory farming is the norm, we're breeding, imprisoning and torturing other creatures on a horrific scale. To take a single species as an example, around 69 billion chickens are slaughtered and consumed by humans every year and most live lives characterised by omnipresent suffering and pain.[2] Of course, those are just the broiler chickens; there are around 7 billion laying hens living in horrendous conditions too.[3] In 1920, the English Professor of Divinity William Ralph Inge expressed concern at the mistreatment of animals in a way that is just as resonant today, declaring:

> … we have enslaved the rest of the animal creation, and
> have treated our distant cousins in fur and feathers so
> badly that beyond doubt, if they were able to formulate
> a religion, they would depict the Devil in human form.[4]

Ok, so we don't get a gold star for our treatment of animals. But we're juggling the issue of feeding a growing global population at a time when developing countries are getting richer and adopting the caloric intakes and dietary expectations of rich countries – which you can hardly blame them for. This doesn't make us evil, does it? I don't think it does, though it definitely makes us the morally culpable architects and enablers of mass suffering. I guess it's a semantic distinction, because we don't *mean* to be evil.

Still, rich countries could easily lead the move towards meat-minimalist diets, house animals in more tolerable conditions, and slaughter them more humanely, and governments could incentivise the accelerated development of synthetic lab-grown meats, which would be entirely cruelty-free. Why don't we? Because we're used to the status quo and we don't want to pay a cent more to ensure that the pigs and chickens we eat don't spend their lives in cages too small for them to move, covered in their own faeces, pumped full of antibiotics, their beaks partially snapped off to stop them pecking at their feathers and cannibalising each other. Got to keep those in the factory farming business employed and those grocery bills down!

For the record, I'm a lapsed pescatarian waiting for synthetic meat to go mainstream. I participate in this horror show as much as the next person. I like meat, life's busy and it's all too easy to go along with travesties that are hard to justify, *even though I know they're wrong*. Like many people in the developed world, I'm rich enough to opt out of this system. But I'm a dumb, selfish ape who knows better and still makes the wrong decision. That's why our ape-brains and their many tragic proclivities need to change.

But perhaps you're not convinced these practices are wrong. Or maybe you think that man's inhumanity to animals is a reasonable price to pay for modern comforts and convenience. Then what of the way we treat each other?

Man's inhumanity to man

We are the species that even now subjugates and enslaves each other by the millions. A species still engaged in civil wars and sectarian violence, arguing over scraps of land in the desert and warring over whose imaginary friend is the one true magic man in the sky. We are the species who continues to do things as insane as removing the external clitorises of female children, or sewing their vaginal openings closed to prevent them having, or enjoying, sex.

None of this nullifies the cognitive psychologist and linguist Steven Pinker's thesis in *The Better Angels of Our Nature*: that this is the best and most prosperous time in human history to be alive. The world really is richer and less violent than ever. We've made staggering progress and the scientific revolution, the Enlightenment, the rise of democracy and individualism, globalisation, and the embarrassment of riches enabled by the industrial revolution have got us here. These forms of humanistic cultural progress have put us on the best footing we've ever been on to make even more progress and eliminate the remaining forms of suffering in our world.

But the great civilisation process that humanity has been through over the past few hundred years doesn't make us intrinsically civilised. We have indeed brought out the better angels of our nature by generating more wealth, and reducing the number of zero–sum games in our world. This takes the pressure off us to act in a dog-eat-dog fashion, but that impulse will return (and still persists in many places) the minute the chips are down and progress stagnates or is reversed. That's why the next step in our civilising process must be to change our animal programming by tweaking our biological natures and upgrading our bodies and minds.

Be honest, you'd upgrade if you could

When it comes to our bodies and minds, we're spectacularly good at doublethink. In the same breath in which we stubbornly romanticise the many confines, wards and dungeons of our biological meatsacks, we'll breathlessly pursue every available opportunity to stave off their decline, upgrade our capabilities, and cast off the shackles of biological determinism.

Who actually likes getting wrinkles? Seeing those first few grey hairs crop up? Wondering if you'd always had that second chin? The answer is nobody. How about menstruation and menopause, any fans in the house? Erectile dysfunction, any big devotees? And that's just the little stuff. How about receiving a terminal cancer diagnosis, or watching an elderly parent with dementia slowly forget who you are, grow fearful, regress into a childlike state, and be medicated into oblivion? Given a choice, wouldn't you do everything you could to prevent or alter those realities?

Our actions and purchasing decisions tell us that we would, and that we do. A growing proportion of people now have the means and the enthusiasm to track their heart rates and dietary intakes, guzzle fistfuls of supplements, have their DNA sequenced, and shoot vials of dermal filler into their faces in order to achieve a simulacrum of the look, if not the essence, of youth. The Harvard Medical School graduate John David made a terrific quip about this in a Facebook post in 2018, writing: 'The anti-aging market is worth $100 billion by selling things that don't work. How much will it be worth when products do work?'

Many of us have spent money on all kinds of products that promise to make us healthier. And sure, maybe we're just drinking kombucha because we love the taste. Maybe we only take supplements because a doctor said it would help with our eczema. And maybe we just love the smell or the feel of that anti-wrinkle cream (we're not really trying to look younger or any silly nonsense like that). Whatever the reasons, these actions help to shift the goalposts of modern health and beauty standards, and our expectations regarding healthcare, quality of life and the optimal lifespan and healthspan are getting higher.

In our own small ways, we're all constantly waging a war against senescence. Let's take the rather superficial example of women dyeing their hair. Most men and women start going grey in their thirties, but how often do you see a woman under the age of 65 with flecks of natural grey? Women don't value having nice, shiny, non-grey hair because they're airheads; they pursue this ideal because it prevents their value from subconsciously decreasing in the eyes of others. It's an imperfect trick. A great hairdo can't make a 50-year-old look 25 again – nothing can just yet, unfortunately. But that doesn't stop the inexorable march to try and regain every last shred of the appearance of youth and vitality.

Evolutionary psychologists have long suggested that long, healthy-looking hair is a signal of youth and fertility. Youth and fertility are important forms of social and sexual currency for women – more so than for men, though they matter for men too. Hence it's no surprise that women have been estimated to spend about twice as much money as men do (and considerably more time) maintaining their looks.[5] Tempting as it might be, you can't lay all the blame for this on cultural messaging and beauty magazines. Women intrinsically care more about their looks because it's more closely tied to our social and reproductive success.

That's a shitty and expensive form of programming to have in you and one that some feminists have been railing against for decades. They just get the targets wrong. Men, the patriarchy, beauty magazines and advertising are proximate, not ultimate causes of women haemorrhaging time and money to primp and preen themselves in an ever-escalating beauty arms race. The ultimate cause is human biology and our hardwired suite of survival and reproductive strategies.

There are evolutionary advantages to using products to enhance your looks, so a large proportion of women will keep pursuing those advantages for as long as our ancient programming persists. If we all decided not to use cosmetics, some subversive clever clogs would instantly figure out that they had a whopping advantage if they bucked the trend. This would be noticed by others, many of whom would copy the strategy, locking us into a vicious cycle. Beauty magazines and advertising are merely symptoms of the evolved impulses that draw us to these vacuous time-sinks in the first place.

Now imagine if there was a pill or an injection that stopped your hair going grey. Or that ironed out wrinkles, prevented skin from sagging, eyesight from fading, sexual function from declining, organs from wearing out, and all those aches and pains and chronic illness creeping in. Would most of us say, aw no, best not mess with the natural order of things, better to age gracefully? Of course not. We'd be begging the companies offering these therapies to take our money, just as we currently throw big dollars at band-aid solutions like hair dye, spectacles, botox, cosmetics, invasive surgical 'mummy makeovers', and Viagra – which mask, but don't actually reverse, the symptoms of ageing.

There are all sorts of things about being human, and being *us*, with our exact set of looks, medical conditions, genes and abilities, that we will do everything in our power to escape. It's just that for most of human history we didn't have very powerful tools or technologies to overcome the worst indignities of our biological destiny, so we rationalised our suffering away and took solace in our children as the torchbearers of the future.

We didn't do this because we prefer to age and die and be replaced by new generations. We accepted it because we didn't have a choice. But the more our choices have expanded and the more our technologies have enabled us to mask or reverse the symptoms of ageing, the more we've adopted them with gusto. When we reach a point where we can be more than human, and cast off our biological limitations and the curse of senescence, we will eagerly adopt those interventions too.

Arguments for and against dying

When anyone tells me that their inevitable death doesn't bother them, I don't believe them. I believe that they believe it consciously, and I'm even a little envious that they *can* believe it. There is a lot to be said for having the wisdom to accept the things you cannot change. But how can we ever hope to change some of the bleakest realities of the human condition if we continue to romanticise them, rationalise them away, or pretend they don't exist?

One example of a thinker who believes humanity is better for the fact that we suffer, age and die is the Professor of Jewish intellectual history Hava Tirosh-Samuelson. She championed this view when contemplating the prospect of treating Parkinson's disease with brain–machine interfaces that could also be used to alter and improve our cognitive capabilities, declaring, 'it is this spectre of transhumanism that makes me most uneasy because it ignores the value of insecurity, anxiety, and uncertainty, which are very much part of being human'. She went on to affirm the benefits of suffering because 'human culture (especially art and philosophy) could not have been possible without these allegedly negative aspects of being human'.[6]

Ah, yes, so it's good that we're anxious, insecure and suffer because we use those experiences to create poems and paintings that help console us. But would we need those forms of consolation if we didn't suffer in the first place? I don't know about you, but if Wilfred Owen's poetry, in which he recalls men 'bent double, like beggars under old sacks, Knock-kneed, coughing like hags', was never written because the event that inspired it, World War One, never occurred, I think that would be a very good trade.

With the loss of World War Two we would probably lose much of WH Auden and poems like 'O What Is That Sound' and 'Refugee Blues'. And in a world devoid of suffering and death, elegies like 'Funeral Blues' would become redundant. So would Alfred Lord Tennyson's musings in 'In Memoriam' that poetry is a 'sad mechanic exercise, Like dull narcotics numbing pain'. And if he had not been faced with the prospect of a premature death in his twenties, John Keats would never have needed to declare for our vicarious pleasure that 'life must be undergone, and I certainly derive a consolation from the thought of writing one or two more Poems before it ceases'.[7]

Every day, roughly 100 000 people around the globe die from age-related causes. That's about one person every minute.[8] What does society gain when experienced people with fully formed adult brains and decades of valuable life experience start losing their mental faculties and physical strength, and spend the next few decades increasingly reliant on medical care, until they cease functioning altogether and we stick them in an urn on a mantlepiece?

The standard shtick is that new generations bring about a reset in culture. We need the old to give way so that the young can break through the stagnant aspects of the status quo, innovate, and move society and humanity forward. Have you met infants, children and teenagers? For all their nice qualities they're hardly a like-for-like replacement for an adult. There's a reason we don't let them drive, vote or run countries. Claiming that death is important so that new life can emerge is obviously a rationalisation for the biological status quo, which at every point in human history has locked us into a compressed cycle of birth, procreation and death. We've had no choice but to make the best of this reality, and so we say, yes, it's sad that people die, but it's wonderful that children are born.

I was surprised to hear Elon Musk say in a 2019 debate with Alibaba's former chairman, Jack Ma, that he thinks 'it's probably a good thing that we do eventually die'. Thinking out loud, Musk pondered the notion with reference to the well-worn quip about the field of theoretical physics: that it advances one funeral at a time. He trailed off with the thought that 'maybe, you know, it's good to have this life cycle'.[9] I wouldn't hold him too rigidly to this position as it was a kind of throwaway musing, and he was questioning it rather than declaring it to be true. But even if it is true that nothing ground-breaking gets done until old fuddy-duddies step aside and make way for progress, it still doesn't follow that they need to drop dead for a cultural reset to take place.

Why is the continuity of selfish genes amalgamated into new individuals who take twenty years to fully form (and another twenty to become really accomplished and interesting) superior to the continuity of selfish genes in the same individual, who can then tweak their genes and undergo such a profound array of cognitive and bodily enhancements that they are rebirthing themselves over time? All without the prolonged neotenous period of having to learn to walk, talk and control their bowels.

If novelty and breaking through the status quo are the objectives, they could be achieved just as well, in theory, by upgrading an existing human and continuing to upgrade them over time. If the ageing process could be slowed or reversed, the knowledge and abilities of older people could be put to much greater use for longer and could accumulate faster, with a little help from our AI friends. I think this is a future we should aspire to bring into being.

Nick Bostrom has also suggested that in a world where more people 'can look forward to a longer, healthy, active life, they will have a personal stake in the future and will hopefully be more concerned about the long-term consequences of their actions'.[10] For anyone worried about climate change, geopolitical stability, nuclear proliferation, or any other big-picture issue or existential risk facing our species, this ought to be a rousing thought. It's a compelling reason to embrace and fund life-extension endeavours, with the aim of enabling people to enjoy vital and healthy lives well into their eighties, nineties and beyond. We'll explore what to do about demographics and the economy in a world of life extension in the second half of this book. We'll also explore some emerging avenues for extending human lives and healthspans in chapter 8: 'Live forever or die trying'.

Torn between two worlds

At present, most people don't think it will be possible to push back ageing and death substantially in their lifetimes. As a result of this belief, we instinctively feel that it would be too dangerous to dismantle our carefully constructed defence mechanisms and let go of our consoling myths about humanity, and so we remain perched on our pedestal. We intuit that relinquishing these props and protective armour could set us up for a blow that would be too devastating to sustain. Unconscious though this decision making largely is, it's also pretty sound and rational at the level of the individual.

If you're reading this, you were born. You have aged since then. You might have watched loved ones die already. And it's more likely than not that you will die yourself before the century's end. The only way we can cope with this is to turn the harsh reality into a consoling story, in which, instead of being erased, we play an important part. Yes, we are going to die, but our families, our children, and the human race lives on. And if you're religious, you might believe we will also live on in the afterlife.

These stories have been necessary coping mechanisms in a world filled with harsh realities. They have not only allowed individuals to avoid being paralysed by existential dread, they have also fostered tribal cohesion, bringing families and groups together for the common purpose of doing their bit to keep the circle turning. They have kept us grounded, and we've needed that. But we cannot stay grounded in these myths any more. Why? Precisely because they keep us on the ground, tethered to the Earth in a circular mode of existence that perpetuates itself *ad infinitum*. A mode of existence that is not going anywhere but down and out.

We need to be brave enough to start supporting the next great human endeavour of life extension, disease elimination and upgrading our intelligence. Even if it doesn't arrive early enough for us as individuals, we should be championing the ideal of a future where my nonna's generation will be among the last to end their days in nursing homes forgetting who they are, shouting fearfully at their children whose names they have forgotten and don't recognise, while wistfully dreaming of their childhoods.

To do that, we'll need to look our mortality and our biological nature deep in the eyes, and find more honest ways to come to terms with them as we cast off our remaining biological shackles, and push the Grim Reaper further towards the horizon.

4

THE GENETIC LOTTERY

I'll never understand what possessed my mother to put her faith in God's hands, rather than her local geneticist.

'Vincent Freeman', *Gattaca*

If scientists don't play God ... who else is going to?

James Watson, quoted in the *Guardian*

The co-discoverer of the structure of DNA, James Watson, regularly gets into spots of bother for his many controversial declarations. Speaking about the possibility of genetically modifying humans, he remarked, 'people say it would be terrible if we made all girls pretty ... I think it would be great'.[1] What's interesting is how much this idea seems to horrify people, particularly those who have long argued that it's an affront to women to judge them based on their looks. I'm afraid it's inevitable that as long as we remain human meatsacks with ancient biological programming, all people will be judged on how they look, in addition to a whole host of other cues about their intelligence, capabilities and ability to garner social status. If you want someone to blame for that, you'll have to chase down a blind architect who designed the whole thing unconsciously: one that goes by the name of evolution.

This kind of admission can make people feel uncomfortable because evolution's not the kind of rapscallion we can easily target and vilify. It's nebulous, intangible, and for better or worse, we owe everything we are to this blind architect. The idea of biology having such profound influence over who we are and how we behave also challenges our intuitive belief that we are rational actors with free will. In reality, we are products of our genes and our environment. Both are forms of happenstance that we were born into, which strongly determine what we are capable of and how our lives will unfold.

Looks matter, we just *wish* they didn't

Watson wasn't being a troll when he said he thought we should make all girls pretty. He was thinking pragmatically, while also engaging in that cardinal scientific sin – having a sense of humour. In rich Western nations we love to endorse maxims like 'true beauty is on the inside' because it wins us cultural brownie points for compassion. But can you honestly say you'd *choose* to have a cleft palate, a gummy smile or alopecia? Or that you've never invested time and money trying to enhance your looks? Having a beautiful mind and an endearing personality definitely count for plenty (and we can thank our genes for much of that too), but you're lying to yourself if you don't think it's an advantage in life to be good looking. We really do judge what's on the inside more favourably if we like what we see on the outside.

It's an open secret that we give tall men more promotions than short men and that men of average height and above have more sexual partners than very short men. Lest we default to comforting adages of quality over quantity, men who are shorter than average have partners who are 'more likely to be less healthy, have lower incomes and education and have higher body mass index'.[2] Simply put, their options tend to be more limited and they have to work harder to prove their worth as high-status partners and bosses. That's often because of some letters they drew in a genetic lottery, as height is a strongly heritable trait. While it's also affected by malnutrition in parts of the world, that's simply a different form of inheritance and the same arguments apply for all forms of inherited luck throughout this chapter – we should attempt to game the system and avoid bad dice rolls wherever we can.

Meanwhile, obese women, more so than their male counterparts, are frequently perceived as being less competent, are stereotyped as lazy, earn less, are considered less attractive by men, have lower rates of marriage, and are less likely to be hired and promoted than their slimmer peers.[3] Of course, there's a correlation with lower socioeconomic backgrounds and education levels in the cases of both shortness and obesity, which could explain things like lower wages, but there's clearly implicit bias in play too. Yes, *cultural* bias – possibly fuelled by unrealistic beauty standards and the overrepresentation of tall men and thin women as leading men and women in TV shows. There's no denying that culture can amplify bias, but where do you think the roots of these cultural biases lie?

When it comes to underlying fitness and fertility cues like facial symmetry, female waist-to-hip ratios, youthful skin and hair, and male muscularity and height (or sexually dimorphic stature), there is strong agreement within and across cultures about what's attractive, and there's not much we can do to outthink it.[4] It's extremely rare for any culture to aspire to morbid obesity, or facial asymmetry, as a normative aesthetic ideal. I know of no examples of the latter, and the only example of the former I'm aware of is the practice of 'bride-fattening' or *leblouh* in rural parts of Mauritania in North-West Africa.[5] And before you say, 'what about the painter Rubens?', his subjects were not morbidly obese.

It is very rare for humans to perceive morbid obesity, or facial asymmetry, as a virtue. Our brains are wired to perceive those traits less favourably because they're honest predictors of ill health – and whether we admit it or not, we *do* perceive them as being less desirable in most contexts – overwhelmingly so in the societies where we are most free to formulate and express our own preferences. For that to change, and for those preferences to scale and persist across time periods and cultures, they'd have to confer other evolutionary benefits that outweighed their costs. That's unlikely, as there are too many alternative, and less costly, ways to signal wealth and status. In most contexts, learning to love those traits and value them for their own sake through cultural reconditioning simply won't work.[6]

Meanwhile, did you know that parents often favour offspring who display signs of having a superior genetic endowment? Sometimes it's looks, sometimes it's smarts, or something else. This makes evolutionary sense – invest your limited resources in the children who are likely to survive and get their genes into future generations.[7] Yet by our modern ideals and values it sounds pretty awful. What kind of horrible person wouldn't love and invest in their children equally, or in proportion to their individual needs? A human one, apparently. I'm sure it's often unconscious. Most of us in the West don't appear to believe it's good to treat attractive, smart, creative and charismatic people as if they belong to a higher caste. But we all do this to some degree. There's even a kind of self-organisation principle in play.

Smarter people often get higher grades, get into better schools, end up with more prestigious jobs and enjoy higher lifetime earnings. Charismatic people find themselves in the spotlight more often than others, where they can garner positive attention and status. The most attractive people often leverage their looks in their careers (as well as their dating and mating careers) and are able to pursue high-status professions like modelling where their genetic endowment is directly commodified. And plenty of average-looking musicians have managed to attract a gaggle of groupies in awe of their lyrics and expressive powers (or perhaps more simply their ability to use those expressive powers to garner fame, wealth and status).

I'll never forget sitting in my GP's office years ago with my mum and dad. The waiting room was packed and there were two toddlers from different families gadding about, circling each other. I forget everything about one of them. The other was mesmerising. She had a wonderfully expressive face that exuded intelligence and self-awareness. She was pushing her pram around the room, toddling up to people, smiling, laughing, clapping and performing – but not in a way I've ever seen a kid play before. There was such depth, engagement and joy to the performance. It was intoxicating to watch. The entire waiting room opted in to the gig, unable to take their eyes off her. I often find people's kids in public places loud and annoying. But I would have gone out of my way to hang out with this kid, would have delighted in seeing her learn and grow, and invested time and resources in her if her family was in my social circle, because she just felt so intrinsically likeable.

Inheriting traits society values in greater abundance doesn't guarantee you'll live a happy or successful life (and there's no denying that socioeconomic background, educational opportunities and upbringing play a role in life success too), but they increase your chances of enjoying more status, wealth, health and access to desirable sexual partners. That's a very big deal for members of a sexually reproducing species whose lives as individuals are, to this day, strongly shaped by our hardwired goals of surviving and reproducing. Status, wealth and high standing in a community help you to achieve both ends. Even if you've had a tubal ligation or a vasectomy and have no intention of *actually* reproducing, you're still programmed to act in ways that would help you fulfil that goal – which is why you probably still care about how you look, how others perceive you and whether or not you're able to attract desirable partners and friends.

Blame it on biology

A partner was telling me about his experiences on Tinder recently, and marvelled at how many women listed a minimum height requirement on their profiles and instructed men to swipe left (left means 'no thanks') if the man was too short. Dating apps didn't create this 'show us your height credentials' trend. They've simply rendered a pre-existing female preference markedly more visible than it used to be, which has unsettled some people for how brazen a signal it is that we're still animals pursuing Palaeolithic-era fitness cues in a modern industrialised world.

When sizing up potential partners in real life, women used to quietly – and for the most part subconsciously – perform calculations like 'is he tall enough?' in our heads. Then we'd make up fake boyfriends and excuses not to go on dates with guys if our inner algorithm didn't ding with 'meets my minimum threshold for physical and other requirements'. Of course what we'd consciously think is something along the lines of 'hmm … I'm just not *feeling* it'. But when matching with partners online, we can't be so opaque, as height's hard to ascertain from a few carefully curated photos – in which short men have a strong incentive to conceal or misrepresent their stature. The fact that so many women have taken to asking for that information upfront and listing it as a dealbreaker (to the extent that some apps now force men to list their height) is a sign of *how much* heterosexual women care about the height of their male partners. If it's enough to risk social disapproval and accusations of superficiality, it must be a lot.

For the record, I think height's a really stupid dealbreaker. But then I've never dated a guy who was shorter than me. Not once, ever. The 'chemistry's' just never been there. Coincidence? I doubt it. I have been with men who were only marginally taller than me though, which may be revealing. Two interrelated things seem to matter for a heterosexual man height-wise in the dating world. The first is being over a minimum threshold of height that most women perceive as acceptable. In Western countries like Australia, that threshold appears to be around the average male height of 5' 8". But that threshold is probably determined by the second factor, which is sexually dimorphic stature, or how tall a man is relative to a woman.

Across cultures and time periods, women prefer men who are taller than themselves. So as a 5' 6" guy, you should be visible to a 5' girl (though probably still less visible than a taller man, because height remains a social dominance cue for men). But you're going to have a much harder time getting the attention of a girl who's 5' 8". By the way, men also prefer women who are shorter than them – women who are shorter than average have the most reproductive success, while those who are 6' or taller receive markedly less attention from men on dating apps.[8] Isn't this an antiquated set of preferences in a modern world where sex and procreation are strongly decoupled, where women can work and support themselves and don't need mates who can intimidate sexual rivals and assert their dominance? Yes! But it's a feature-turned-bug that we're stuck with until we seize the evolutionary reins and give humans the tools to start choosing their own genetic endowments and cognitive preferences.

Unfortunately, we're all the dupes of biology in one way or another. Throughout history, we've pushed back against our inherited fetters in any small way we could. It's time to take things further and intervene at every level of our biological design: genes, cells, cognition and the evolutionary motivations that underpin our existence. Until we do, many more people will continue to suffer as a result of the innate human propensity to judge certain traits less favourably. If nobody was born with a predisposition to obesity, or with crooked teeth, noses or spines, and all men exceeded the minimum height thresholds that render them sexually visible to women, we could greatly reduce the suffering that pronounced genetic inequalities engender in our world, both directly and indirectly – to say nothing of the legion inherited inequalities in the domains of intelligence and health, which we'll get to presently.

This *definitely* won't eliminate competition. But it might help shift our focus to higher-order points of differentiation, and perhaps reduce how marked the disparities in life outcomes currently are between the biggest winners and losers of the genetic lottery. Particularly if we are able to enhance human intelligence. There's only so far we can shift cultural ideals, because no cultural norms exist independently of human hardwiring and evolutionary strategies. You might have the most progressive values in the world and harbour a genuine feeling of love and kinship for all humanity. You might believe that all people are truly alike in dignity and vocally defend the rights of everyone to be seen, heard, loved, cherished, desired and represented on magazine covers and in TV shows. You might declare that you would never be less likely to hire, date or have sex with someone who wasn't attractive, tall, slim, smart or young enough. But you will not live by it, not truly.

Worried about diversity?

James Watson's son Rufus was born with a severe form of autism, suffers from schizophrenic symptoms, and was non-verbal until the age of four. He has lived with his parents throughout his adult life when not institutionalised, and is unable to look after himself unaided. This is one of many tragic stories of genetic happenstance, which patently belies some of the more fanciful ideals of diversity that are currently in vogue. It takes all sorts to make a world, but it doesn't take every possible form of diversity to make the *best* world.

It's hard to argue in good faith that it's wonderful that some people are brought into this world with such profound limitations and disabilities. There clearly is value to some autism spectrum traits (and my hunch is that on average, transhumanists possess a higher payload of some of these high-functioning traits). But being unable to function autonomously in the world is not a benefit of any kind – it's a profound tragedy. Watson has said as much about Rufus, publicly stating, 'I use this term genetic injustice when I think of my son'. When the human genome project commenced, Watson looked ahead to a future where there would no longer be 'parents who lived with 30 years of terrible uncertainty while they tried to find out what was wrong with their child'.

There is nothing insidious about acknowledging that it would be better if some traits were no longer heritable. Think about the deleterious BRCA1 and BRCA2 mutations that strongly predispose their carriers to breast, ovarian and prostate cancer. What is the benefit here? There might actually be one that we haven't discovered yet, but who wouldn't want to decouple the beneficial traits from the extremely undesirable ones? Meanwhile, it would undeniably be better if no child was ever born with a congenital heart defect, a missing limb or severe intellectual disabilities.

People who are born with conditions that make life harder are human beings who deserve all the support we can reasonably give them. But *conditions* that severely limit a person's life arc and opportunities are bad, and we should do everything in our power to try and eliminate those conditions. In doing so, we will select against some forms of diversity, because not all forms of diversity are good. Paedophilia and psychopathy are clear cases in point.[9]

I know, that opens up the famous old can of worms: which are the 'good' forms of diversity and who gets to choose? Well, who's choosing now? You are, sort of, in a very haphazard way. For the most part, people in the West choose sexual partners, and we don't choose them by accident. We are subconsciously 'trait selecting' when we decide who to breed with. It's very imprecise and we're relying on imperfect outward cues of status, health, fertility and fitness, and there are always trade-offs. But we are selecting for certain traits and selecting against others when we choose mates. There's nothing insidious here. So why do we think it's insidious if we do it a little more precisely and eliminate many sources of human suffering in the process?

Well, it's a bit like how the Chernobyl disaster put the kibosh on the future of nuclear power, tainting that meme for generations. A long time ago, a militant moustachioed maniac had a hare-brained idea. He wanted to make Germany great again and thought he needed to purge certain types of people from the population to do that – mostly Jewish people, but also 'Gypsies', homosexuals and others. It's an age-old tactic – mobilise a tribal in-group against a pilloried out-group. Hitler capitalised on centuries of anti-Semitism in a bid to unify the rest of Germany under his political regime. In the process, he instigated one of the greatest horrors of human history, in the form of the Holocaust.

In the second half of the 20th century, humanity rightly focused on trying to ensure that such a thing could never happen again. But one of the consequences of our diligent efforts to disavow, and dissociate ourselves from, anything related to national socialism was that we overcorrected in some ways. We demonised ideas and concepts that were not of Nazi origin, but were co-opted and distorted out of all recognition by Hitler and his followers. As such, I think we need to take a moment and try to disentangle our justified digust at the Nazi regime, from a word that's been cleaved to it ever since: eugenics.

Eugenics literally means 'good birth'. As the *Stanford Encyclopedia of Philosophy* notes, 'intuitively we have some moral obligation to promote good births – to have, in the most literal sense, eugenic aims. Indeed, if parents are encouraged to provide the best environment for their children (good nutrition, education, health care, a loving family situation, etc.), why not also encourage them to ensure their children have good genes?'[10] The encyclopedia distinguishes 'authoritative' eugenics, which are coercive – where we are told from on high what sort of people there should be – from 'liberal' eugenics, where individuals are free to choose and can select against debilitating, or fatal, conditions like Huntington's disease.

Arguably, there could not have been a truly benevolent, and highly effective, eugenics policy in the early to mid-20th century, as the human genome was so little understood. And what Hitler enacted in the 1940s was *not* a lucid eugenics policy – it was a pogrom motivated by fear, hate and mis-understanding. There was no sound scientific rationale for his hamfisted endeavour; it was the product of paranoia wrapped in pseudo-scientific packaging. Yet he has single-handedly managed to scare generations of Western scientists, journalists and intellectuals into not mentioning eugenics ever again (*don't mention the war!*). But how can we ever learn from the war if we don't mention it? Or explore the profound benefits of genetic selection in the modern era if we refuse to even say the word eugenics, as if fearing that a lightning bolt will strike us down if we do?

Let's imagine the media's reaction if I were the first woman to have a child that was aesthetically 'designer', rather than simply having genes tweaked to avoid disease. They'd absolutely zero in on choices I'd made regarding sex selection, eye colour, racial traits, markers of intelligence and so on. Fair enough, they're interesting questions. But I don't think many of those choices are as interesting as clickbait articles on designer babies make them sound.

Honestly, who cares if you want a kid with green eyes instead of brown eyes, or curly hair instead of straight? This really is superficial stuff. Let's say everyone in the world decides that people with Latino genes have the most desirable look, and we start selecting for those looks in mass numbers, leading other aesthetics to dwindle. It literally doesn't matter. Wanting the exact proportion of what we think of as racial traits to be retained in the global population in perpetuity is like wanting to preserve every species that currently lives on Earth, while not caring at all about the fate of the 99 per cent of species that ever lived that have long been extinct – or the many future life forms that might supersede the creatures on Earth today. It's not diversity we're championing, it's the status quo and its familiar tribal identities, which are anything but helpful or globally cohesive.

To address the scary image of us all becoming carbon copies, I don't think this would happen, and people who imagine it would display a spectacular failure of imagination. We're more likely to invent new aesthetic ideals based on the recombination of many traits, amplifying the already strong trend towards mixed-race heritage and aesthetics.[11] Eventually, as we move away from biological bodies, we'll move away from two-armed, two-legged meatsacks altogether and get much more creative with our avatars. The only constant in any evolutionary system is change – we'll simply be exerting some new selection pressures, just as we did when we changed every domesticable crop and animal on the planet. Are our labradors and French bulldogs wrong because we *selected* for them at the expense of other forms of canine diversity?

Remember, whatever physical traits you have now, they are already the product of migration, colonisation, selection and recombination. You might very well associate your identity with having straight hair, blue eyes, or dark skin, and have perhaps come to associate that image with a particular national flag (a flag that is itself no more than a few hundred years old). If you imagine any familiar set of traits that matches up with an existing racial or cultural identity is going to continue to exist in large numbers in a century or two, you're mistaken. More importantly, why would we *want* those forms of identity to persist?

Once upon a time we were all African. We've diverged as we migrated across the globe, expanding into new niches that exerted different selection pressures on different tribes. Lighter skin was selected for among those who ended up in the northern hemisphere away from the equator, where there was less ultraviolet radiation, allowing more UV photons to be absorbed through the skin to produce adequate levels of vitamin D. Darker skin was selected for where UV levels are higher, and is protective against skin cancer and sun damage.[12]

Neither dark nor light skin is inherently superior; they are adaptations that were useful when we had no other ways of getting our vitamin D, or reliably protecting ourselves from the sun. It would have been much more useful for me, growing up in Australia, not to have had Irish skin, which is particularly thin and translucent. I sometimes think it's tantamount to having a disability – I spend so much effort covering up and wearing sunscreen, and I still seem to get burnt and freckled after more than ten minutes in the sun.

If it makes you uncomfortable that I just equated my pale skin with a disability, remember that Australia and New Zealand have the highest rates of skin cancer in the world and my susceptibility to melanoma might actually kill me at a premature age. I'm not well evolved for life on that continent and I'd change my skin type in a heartbeat if I could because it's ridiculously impractical. If that means selecting against Irish DNA, so be it.

We're also selecting against racial homogeneity all the time anyway, with record high levels of interracial marriage. Some people worry that we might reverse this trend, revert to an Aryan ideal and all choose to have children with blond hair, pale skin and blue eyes. I think this is a really misplaced worry. For starters, global beauty standards are already converging on a mixed-race aesthetic. The more morphological freedom technology bestows upon us, the more we're going to be excited to explore new combinations of traits – and eventually the racial categories we're familiar with today will fall away. In the meantime, it's worth remembering that there are much more interesting, useful and relevant traits to think about when selecting genes than what colour stuff is. It's time to stop judging others by the colour of their skin when what really matters is the content of their character.

How do you rig the lottery?

It's a profoundly transhuman world when the most fundamental components of who we are, the As, Ts, Cs and Gs that make up our genome, are fair game for altering and re-writing. But is it really that easy to change some letters and transform a person? In practice, no, not yet. In no small part because most traits aren't Mendelian, or determined by just one gene – they are polygenic, which means they are influenced by many genes (often hundreds or thousands of them). But as branches of information technology, genomics and its medical discovery accelerant, AI, are both becoming exponentially more powerful and cost-effective as computation (in terms of both hardware and software) continues to get cheaper and more efficient.

More computing power enables more data to be processed and interpreted faster at lower cost, and data-mining algorithms can devour big datasets to detect new patterns, which in turn accelerates new discoveries. We're poised to see an exponential cascade of new knowledge about the human genome and the function of many genes this decade, which will refine and reshape our map of human biology and human nature – including the way we think about personality traits, disease, behavioural causation, health and medicine.[13]

What's missing right now are the big genomic datasets. Most people today, even in the world's richest countries, have not had their entire genome sequenced. So how do we know that rapid user adoption is likely in the coming years? Because whole genome sequencing is just about cheap enough now to become a democratised technology. We can see that evidenced by the rise in user adoption in recent years – both through direct-to-consumer tests offering ancestry information from partial genome sequencing, and the increasing use of whole genomic sequencing in public healthcare systems in the US, UK and elsewhere.

In 2012, the UK government announced an ambitious initiative called the 100 000 genomes project. Patients with rare diseases and cancer were recruited to have their whole genome sequenced, with the aim of catalysing discovery and innovation in genomics and healthcare, and 'providing the foundation for a new era of personalised medicine'.[14] The 100 000th genome in the project was sequenced in 2018.[15]

In late 2018, the UK government announced a subsequent plan to sequence five million whole genomes between 2019 and 2024. Since 2019, all adults with hard-to-treat cancers and certain rare diseases, as well as all seriously ill children, have been offered whole genomic sequencing as part of their care in the UK. Meanwhile, in the United States, the National Institutes of Health (NIH) have allocated nearly $30 million to establish three genome centres, which are building a database of genomic and lifestyle information that will be garnered from at least one million research participants.[16]

Humanity's big genomic dataset is growing fast. As it does, AI continues to get more powerful and the cost of genomic sequencing continues to decline. That's why we can be confident that a deluge of insight is coming in a blink-and-bam timeframe. The cost of sequencing human genomes has famously declined *faster* than Moore's Law, which means we could be on the cusp of sequencing most genomes on the planet for next to nothing.[17]

Imagine you're born in the US in 2030. The midwife tells your parents you look perfect – an ideal weight, with ten fingers and ten toes. You're a happy, healthy, thriving baby. A drop of blood is taken in the first moments of your life and your genome is sequenced on the spot. From day one it's known that you're free from rare genetic disorders. That genomic data is stored on your Electronic Medical Record (EMR) and will be cross-referenced in every medical consult you have throughout your life.

The data becomes more useful every year you live as millions more babies are born, genomes sequenced, and insights unearthed. When you're in primary school it's discovered that you're likely to have difficulty metabolising certain vitamins and minerals and need extra supplementation. Your sibling is told they are likely to have an adverse reaction to certain forms of anaesthesia, but you don't have the same genetic markers and can use those drugs safely. As more data pours in over the years, you learn that an antidepressant that works great for some people might increase suicidality in people with your genetic profile.

Your genome will also yield ever more robust predictions about your intelligence, personality traits and abilities from birth. A lot of people don't like the sound of that. But our genes are predictive of all those things now, we just don't have the tools to read the map ultra-precisely. We should be actively striving to develop those tools, and more computing power and AI are very likely to help accelerate that endeavour. Robust knowledge about what makes people who they are won't lead to the genetic determinist society of *Gattaca*; it will give us the foundational knowledge we need to change and improve our genetic destiny.

This isn't science fiction, it's imminent. In the US today, newborns already have blood drawn at birth and are tested for a select number of conditions, including cystic fibrosis and sickle cell anaemia. One in 300 babies screened will test positive for a condition on the panel.[18] But those with other kinds of heritable diseases remain undiagnosed for years, because we don't have the comprehensive genetic data to diagnose them on the spot. Ubiquitous whole genomic sequencing could change all that. It has the potential to dramatically reduce the number of infants born with heritable diseases, as adults of reproductive age will have their whole genomes sequenced and compared for genetic risk factors before trying to conceive.

It's funny how you can be living on the cusp of a major transformation and not realise it until you blink and find yourself on the other side. If you're wondering why I've spent much of this chapter talking about the theoretical benefits of rigging the genetic lottery, think back to the nature of exponential trends we discussed in chapter one. Should we explore the promise and peril of personalised medicine and gene editing technologies in the earlier stages of their development, while we're still on the first half of the chessboard? Or should we wait until the exponential decline in price puts these services on to an accessible app and overhauls healthcare as we know it?

Knowledge is power

For decades, humans have debated whether it's right to attempt to play God and rig the genetic lottery. The answer is: definitely! I don't know about you, but given the choice I'd prefer to have been designed by a *conscious* architect who could see, and who had a clear goal in mind when designing every bit of me, gene-for-gene, as opposed to a blind evolutionary designer making a minimum viable product, tweaked from older blueprints (like a Boeing 737-MAX 8 jet) so that it might just be solid enough to survive and reproduce if we cross our fingers really tight (pretty sure the MAX 8 planes won't survive, but you get the idea).

I've never had my whole genome sequenced, so I haven't got a clue what ticking time bombs lie in wait for me. A genetic counsellor couldn't tell me much at this stage anyway as we don't know what most genes in the human genome are for (we're still waiting on those big datasets). I have had my partial genome sequenced though and I don't have the deleterious BRCA1 or 2 mutations. That doesn't come as a huge surprise as those mutations are rare, and I have no known history of breast, ovarian or prostate cancer on either side of my family. But what of the other 20000–30000 protein-coding genes inside me?

How likely am I to get other cancers? Heart disease? Wind up with a serious mental illness? Suffer from Alzheimer's or Parkinson's disease? Whatever my predisposition to these and other conditions it's already largely written. Of course, the outcome of my life isn't written. Having a 50 per cent lifetime risk of contracting Parkinson's disease doesn't mean I'll get it and there could be many lifestyle modifications, environmental triggers or epigenetic switches that could edge things in one direction or the other over the course of my life.

That's precisely why I want to know. Knowledge is power, and human genomes represent a huge untapped repository of future knowledge that we could use to ameliorate human suffering and radically improve our lives and those of generations to come. And not just for living people like you and me, who are the walking products of random dice rolls.

Imagine if the dice could have been loaded favourably at conception. We could live in a world where no mother would ever again give birth to a perfect looking child, only to find out the child has a debilitating or life-threatening condition. A world where you don't ever have to scroll through Facebook and see a post from someone you know announcing with a brave face that they're battling cancer. Such ticking time bombs are part of our state of nature, and it's time we blasted them out of existence. We have the tools to begin embarking on this endeavour today.

With the help of technologies like AI, we stand a chance of doubling down on what we've been doing throughout human history: extending our life and healthspans, un-shackling ourselves from pain, suffering and degradation, and cultivating much more ambitious ideals of what a good life looks like. There are better possible futures where we no longer have to care what someone looks like, because our hardwired preferences have been rewritten. There are futures where what truly matters is intelligence, rationality and empathy, and where our capacity for these things is superhuman.

Don't forget, it was also James Watson who said, 'I am sure that the capacity to love is inscribed in our DNA'. He reflected that 'if some day those particular genes, too, could be enhanced by our science to defeat petty hatreds and violence, in what sense would our humanity be diminished?'[19] In what sense, indeed. The best version of humanity is one that honestly examines its flaws and limitations and seeks to diligently overcome them.

5

APE-BRAINS IN
A MODERN WORLD

*Two things are infinite: the universe and human stupidity;
and I'm not sure about the universe.*

attributed to Albert Einstein

My brother and I like to play hypotheticals, especially the
'who would you save' trolley problem kind, which often
make our parents squirm when they overhear us weighing
up the value of the lives of family members and pets. This one
really highlights how tribal I am and it's very unflattering
– except towards my brother Dom. I have to imagine there
are two levers in front of me. If I pull the green lever, Dom
will die instantly. If I pull the pink lever, a random stranger
somewhere in the world will die instantly instead. If I do
nothing, both will die. Who do I save?

It's a no-brainer that I rank the life of my brother higher
than a stranger's. But how much higher? Would I allow ten
random strangers to be killed to save his life? Yes, I don't
even have to think about it. What about a hundred random
strangers? I know this is very, very, *very* wrong, but selfishly,
I think I would still save my brother. It's hard to know what
I would actually do if I were sitting in front of a real set of

levers that I had to pull, knowing the consequences weren't imaginary. Maybe my conscience would get the better of me. But emotionally, Dom is worth more to me than a hundred, or even a thousand strangers.

I suspect someone you know is worth more to you than the lives of a hundred strangers too. Congratulations, you're a biased human being with a monkey brain. The great news is that this *exact* scenario doesn't tend to happen in real life, where your family members and strangers are randomly kidnapped and their lives played off against each other. The bad news is that, in a world of global diplomacy, unequal wealth distribution and massive quality of life and opportunity differences between rich and poor nations, this tribal bias comes into play all the time, as the wealthiest countries shore up their borders and citizens demand welfare spending be kept for themselves, while deeming that the poorest of the poor should accept their lot in life and go back to where they came from.

Through our actions and inaction we are very literally deciding who lives and who dies in the world on a regular basis – and in many cases it's not even a case of letting one die to save another, as when a doctor has two pneumonia patients and one ventilator and has to decide if it should go to a 75-year-old smoker, or a previously healthy 25-year-old. We routinely let people die *preventable* deaths across the globe and we rarely bat an eyelid in rich Western nations when people in Africa get Ebola, or malaria, or live without access to HIV medication, vaccines or clean water. We're used to those specific forms of suffering disproportionately occurring in certain parts of the world and they scan in our brains as being unfortunate, but part of the plan.

Not that these are easy problems to solve, and it's not clear how much should be spent, or how resources can be most effectively deployed, particularly in unstable or war-torn countries. But whether it's keeping welfare spending domestic, or wrestling each other in the supermarket aisles over toilet paper and canned food during a pandemic, it's clear that we overwhelmingly prioritise our local, immediate, tribal, familial, domestic and personal needs above those of the collective, because it's how our ape-brains evolved in our small, kin-ordered tribal societies.

If you're biased and you know it clap your hands

Few people in the world are more aware of human cognitive biases than the Israeli–American behavioural economist and Nobel laureate Daniel Kahneman, who has spent a lifetime researching them and communicating his findings. When interviewed by Sam Harris on the *Making Sense* podcast in 2019 and asked if his insights had enabled him to modify his thought processes and behaviours to avoid basic cognitive traps, Kahneman replied, 'not at all'. He noted that he was sometimes able to recognise certain traps and cognitive illusions in the moment, but stated, 'I don't think that I've become in any significant way smarter because of studying errors of cognition'.[1]

Harris seemed surprised and crestfallen upon hearing this. If Danny Kahneman can't consistently be less wrong, what hope do the rest of us have? But what's so wrong about the way humans think anyway? Simply put, our brains haven't evolved to perceive reality as it is. We see the parts of the spectrum of reality that were most useful for surviving and reproducing on the African savannah a hundred thousand years ago. With that in mind, let's have a think about how some of these biases affect us in the modern world.

During the Cuban Missile Crisis and throughout the Cold War, the prospect of the world ending felt very real. But after the fall of the Soviet Union, we relaxed a little, and felt like reason had prevailed. Yet think about what's happened in the meantime. Today, *nine* countries have nuclear weapons, up from only two in 1950. While the number of active stockpiled warheads has declined significantly since the end of the Cold War, and their explosive energy has been reduced, the old superpowers are continuing to upgrade their arsenal, and many warheads remain on hair-trigger alert, which means they can be deployed within minutes.

The risk of a nuclear war by accident or intention is still grave, but we don't think about it day-to-day any more than we think about the fact that one day our planet will become uninhabitable and the sun will die. There's something about this knowledge that gets marked in our brains as 'textbook fact'. The kind of thing you say and recite dispassionately, and rubber stamp with the imprint: FAR FUTURE, OR IMPROBABLE: DON'T WORRY ABOUT IT. It's a remote thing and it's surely not going to happen in our lifetime.

Part of why we're so complacent about the risk of nuclear proliferation and nuclear war is because they're complex global problems that most individuals can't do much about. It also doesn't help us get on with day-to-day living to dwell on scary, grim prospects. But there's another reason, and it's got to do with something called the *availability heuristic*.

When assessing the probability of something happening in the future, we scan our memory banks and look for examples of it happening in the past. How many times has humanity been upended by a nuclear war? None. Phew, I can count the times on no hands! Feels like a pretty low risk.

Most of us don't really believe a nuclear war will ever happen, because our brains falsely tell us, after scanning historical data, that it's super unlikely. In his recent book *The Precipice*, the Oxford philosopher Toby Ord estimates that the risk of a nuclear war causing human extinction in the next hundred years is in the order of 1 in 1000. That's an order of magnitude higher than Ord's estimated risk from a natural pandemic or supervolcano eruption.[2] And that's just the estimate for a war so bad it drives humanity to the brink of unrecoverable collapse. A less catastrophic nuclear war is even more probable.

Add these dangers to the burgeoning suite of other man-made existential risks like bioengineered pathogens and artificial intelligence, and Ord puts the odds of humanity facing extinction in the next hundred years, or undergoing a collapse from which we lose our potential to recover, at one in six – or the roll of a dice. That's if we take mitigating steps now and alter our course. In conversation with Robert Wiblin on the *80,000 Hours* podcast, he put the odds at one in three if we do not.[3]

This is frightening given how bad humans have historically been at proactively changing course and estimating risk. Thanks to the influence of biases like the availability heuristic, we grossly overestimate how dangerous things like flying are, because we've all seen footage of plane crashes on TV multiple times, and hundreds dying in a single event is impactful because that could have been the annihilation of our entire tribe once upon a time.

In the twelve-month period following the 9/11 terrorist attacks on New York City and Washington DC, it's estimated that an additional 1500 to 2300 Americans died in car crashes as a direct result of avoiding air travel and spending longer on the roads.[4] Our inability to accurately assess risk leads us to make poor decisions, while feeling in our gut that they're right – and that puts us in grave danger. On the scale of individual decision making it's not the end of the world. But it can be catastrophic when we do this on the scale of the species.

Humanity keeps failing the marshmallow test

One of the most famous social science experiments of the 20th century is colloquially known as the marshmallow test. In this experiment, a young child is presented with a treat, like a marshmallow. They are told they can eat the treat right away, but if they wait and don't eat it until the researcher comes back into the room after about fifteen minutes, they'll get two treats. If they gobble the marshmallow *before* the researcher returns, they won't get a second one.[5]

With delayed gratification you can double your reward, so it pays to be patient. But as a species, humanity is a lot like a 4-year-old child who just can't help grabbing a marshmallow when it's placed in front of them. We reach for the immediate reward even when we're told we can have an outcome that's *much* better. In fact, we choose short-term gains even when we know that delayed gratification and careful planning for the future can help us avoid dire catastrophes, which would result in us never getting any marshmallows ever again.

With that in mind, let's delve into one of the most topical examples of humanity's inability to forgo short-term gains and prioritise future wellbeing. Until the emergence of the novel coronavirus, SARS-CoV-2 in late 2019, the last time humanity was faced with a major respiratory pandemic was in 1918 with the Spanish flu, which infected one-third of the global population and killed tens of millions of people, all before the age of commercial aviation.

As a global community, we've had a century to up our game since then, and forerunners like the Hong Kong Flu of 1968, and SARS and MERS, should have put the world on high alert. We've had ample time to focus on building robust global health systems, putting tribal and national differences aside, and investing in the future safety of the human collective, rather than prioritising short-term economic and political gains.

What's happened instead? We've put a lot of necessary funding and skill building on the backburner. It's hard to get enough medical graduates into epidemiology because they get paid much better in other fields and they follow the money – or their personal preference for careers that involve more immediate positive feedback and rewards. We don't have a lot of surge capacity in the system in case of a global emergency, and many countries are painfully short of ventilators, trained medical staff to operate them, and personal protective equipment (PPE) like masks, gloves and gowns.

The evolutionary biologist Heather Heying used a great analogy to describe this situation, based on a story about some observations her graduate school professor once made about squirrels. He thought the squirrels were hilariously wasteful for burying far more nuts than they ever dig up in a given year. But as Heying points out, the squirrels were burying the nuts for the worst years that only come along every now and then. Those squirrels who don't bury enough nuts benefit in the short term, but when times are tough, they are at the mercy of those who planned ahead.[6] When it comes to pandemic preparedness, it's clear that humanity hasn't buried enough nuts.

Worse still, even when the pandemic hit, many leaders bungled their response, or concealed important information from their citizens and from global health authorities. China is the first example, with leaders initially attempting to suppress knowledge of the local outbreak and conceal it from the rest of the world. Iran was the next country to make a major gaffe, prioritising positive public sentiment in the lead-up to their imminent national election, over viral containment and the safety of its citizens, which led to a catastrophic spike in infections. In August 2020, data leaked to the BBC revealed that Iran's case numbers and fatalities were also being severely underreported to the world.[7]

On 3 March 2020, the UK's Prime Minister Boris Johnson boasted that he'd been to a hospital with confirmed coronavirus patients and he 'shook hands with everybody'.[8] This was back when he was trying to prevent panic and assure his citizens that everything was a-ok. Within weeks, Johnson was diagnosed with the coronavirus and the UK government had backtracked on its relatively relaxed mitigation strategy. Shortly afterwards he was hospitalised and placed in intensive care. After a brief period of implementing more cautious policies in the aftermath of his recovery, Johnson reverted back to mixed messaging and short-termist policies like 'eat out to help out'. This attempt to reboot the UK's economy before the epidemic was under control necessitated protracted lockdowns and resulted in a subsequent wave of infections and preventable deaths.

When it comes to Donald Trump's response in the United States it's hard to know where to start. There was the early denial that the virus was concerning and posed a more serious threat to Americans than the flu, which was later revealed in a taped interview with the journalist Bob Woodward to be a calculated lie.[9] Then there was the refusal to act throughout the month of February 2020 by promoting physical distancing or making preparations in the event of hospitals being overwhelmed.

In staggeringly tribal fashion, there was also a strong focus on rebranding humanity's common microbial enemy as the 'China Virus', in an attempt to shift blame to a national adversary and create an easy scapegoat for the ramifications of America's political inaction. We can also add the stigmatisation of mask-wearing to the seemingly interminable list, along with the fact that Trump also went on to contract the virus along with the First Lady Melania, their son Baron, and dozens of other staff members in a White House superspreading event.

From India, the writer Suganya Lakshmi reflected on the feeling of 'blind positivity' she and others in her community felt in the early stages of the pandemic. In her words, 'We already lived in pollution, surely our immune systems would be able to beat a "flu".' She hinted at the need to think that way in India, for 'there was absolutely no way a country of that size could ever shut down'. Not when so many citizens are daily wage labourers who can't work from home and must work to survive.

It's true that countries like India were not in a position to eradicate the virus. But even among those more privileged, who could have modified their behaviour and thought more carefully about the prospect of a new normal, or a second wave, complacency took hold. She writes:

> As the year went on, we spoke less and less about COVID-19. We let our guard down, casually meeting friends, removing our masks to beat the heat, forgetting our sanitisers and letting go of all pretence of social distancing. Some of us took advantage of the new WFH lifestyle, leaving big cities to move to the mountains or to the seaside. The social media accounts of India's privileged showed off pictures of exotic holidays in the Maldives and slightly less exotic parties in Goa. Politicians held election rallies and citizens flocked to places of worship.[10]

In March 2021, a second wave hit. Supplies of oxygen were constrained, hospitals were overwhelmed and crematoriums were full. By 1 May, 400 000 cases of Covid were reported in a single day.[11] Lakshmi's father contracted the virus that week and was admitted to the ICU. He died seventeen days later. While it's reasonable to modulate your behaviour as your risk profile changes, this story speaks to something more fundamental about human psychology: the wishful desire to believe everything will go back to normal, and the consequent dismissal of dangers, contingencies and tail-risks that you keep telling yourself won't happen because (again, present bias and the availability heuristic) they haven't happened to you before, and because *you really don't want them to happen.* To someone else, fine, maybe, but not to you.

If you're living hand-to-mouth in a poor country, it's probably not helpful to dwell on emerging risks you have little to no control over. But what you should be able to rely on are leaders who will engage in robust society-wide planning, so that you face as little disruption as possible. When the people in charge refuse to dwell on big global risks, which they should be working tirelessly to try and mitigate for the long haul, we all end up worse off and live through times of precarity for much longer than we should.

Yet in the early stages of the pandemic, we saw our leaders watch a storm approach and do nothing. It got closer and closer, and started to hit some parts of the world, and doctors, nurses and leaders in hard-hit cities begged the rest of us to act now before it was too late. Our leaders took in the evidence, but decided that football matches and short-term economic activity were more important and we could deal with the storm if and when it hit. This is a classic case of *present bias* coming in to play. Unfortunately, humans are apt to give more weight to short-term payoffs when considering trade-offs between the present and the future.

Our leaders also succumbed very hard to the *normalcy bias.* We don't plan for, or appropriately react to, a situation or calamity that has never happened before (this is very similar to the availability heuristic). We plod along as usual, because we're wired to expect a continuation of the status quo and we're reluctant to update our expectations until something literally comes out of nowhere, shakes us up, and forces us to confront a possibility that we weren't expecting.

Perhaps most galling of all, we watched leaders continue to make the same mistakes, and prioritise short-term political point scoring over long-term strategy well into the second year of the pandemic. Making a big mistake in a novel situation is understandable. Making the same kinds of errors over and over is unforgivable.

This pandemic has made it starkly apparent that the traits we select for in our leaders are not traits that make them suitable candidates to ensure human sustainability and prosperity over the long term. We don't tend to revere softly spoken brainiacs with true foresight, and overwhelmingly select leaders who we think of as being charismatic and part of our tribe. We favour those who are good actors and spin doctors, rather than deep thinkers and strategists. I think Douglas Adams put it best when he wrote in *The Restaurant at the End of the Universe*:

> To summarize: it is a well-known fact that those people who most *want* to rule people are, ipso facto, those least suited to do it. To summarize the summary: anyone who is capable of getting themselves made President should on no account be allowed to do the job. To summarize the summary of the summary: people are a problem.[12]

Suckling at the clickbait teat

Another major bug of our ape-brains – which relates to the short-termism we've already talked about – is our propensity to gorge ourselves on sensational misinformation. We've always had the bug, but it used to be adaptive back when a sensational event was likely to be an encounter with a tiger, a rival tribe, or a bad actor in your own tribe. With brains heavily attuned to novelty and drama, you could pass on stories about those events very effectively, conveying useful warnings and moral teachings that would have aided survival.

Nowadays, the opposite is more likely to be true. Our tribes could crumble as a result of this ancient, maladaptive hardwiring. When it comes to clickbait, we know it's brain-junk, we know it's bad for us, and we know we shouldn't just read the headline and then share the meme without fact checking. But who ever let facts get in the way of a good story? Even those of us who have been trained to think carefully about the quality of evidence can't help but look if we come across a claim that is sensational.

I came across a story in March 2020 claiming that the Belgian health minister banned 'non-essential' sex in an indoor setting between three or more people. The article was shared on a group chat I'm in. The first red flag was that the story was posted on a site I'd never heard of called worldnewsdailyreport. Still, my first instinct was to send it straight to a friend of mine, because we'd been discussing (and lamenting) the coronavirus-induced sex hiatus among the many of us who don't have live-in partners. I wanted the story to be true because it was a good conversation prompt and it was kind of funny. But I stopped myself and did a quick fact check. The story was, of course, completely false.

Why was I so drawn to it over, say, a story about the Belgian health minister urging citizens to wash their hands? Because handwashing is banal, while banning group sex is unexpected and evokes intriguing sexual imagery – which is powerful, because humans are susceptible to a bias called the *bizarreness effect*. We are much more likely to remember something novel, sensational or strange than something ordinary. It's why the media overreports on murders, car accidents, plane crashes, fights, riots, shark attacks and other 'if it bleeds, it leads' stories. The glut of these stories gives us a false sense that the world is much more dangerous than it is, and these phenomena keep being reported because we're hopelessly drawn to them.

Our fondness for sensational stories and their ready commodification skews our information diet heavily towards what we've recently come to term 'fake news'. As a collective, we're smart enough to know and recognise this, but we're also seemingly too wedded to the status quo and too swayed (it's the ape-brains again) by the short-term commercial incentives that perpetuate it.

We're also prone to something called the *availability cascade*. We deem things more plausible the more times we hear them repeated in public forums. It's as though our ape-brain says, yes, look, tribal elders are talking about this. No need to look into the evidence, do as the tribe does, say as the tribe says, believe as the tribe believes. Couple this with the *continued influence effect* and the *illusory truth effect* and we end up believing a lot of bullshit, as we tend to continue to believe false claims, even after they've been retracted and corrected. If we hear something enough times, like that face masks don't help contain the spread of respiratory viruses, it ends up having the ring of truth to it.

Humans are also really bad at thinking in terms of probability, which is a shame, because all forecasts are estimates, from the weather, to the likelihood of a plane crash, to the odds of humanity going extinct from a nuclear war. If, in a hundred years, no nuclear war happens, that doesn't mean the odds of it happening over the course of the century were 0 per cent. Just as, if you're a very unlucky person who happens to be in a plane crash, the odds of your plane crashing when you boarded the flight weren't 100 per cent; they were more like one in several million.

But because of our susceptibility to *overconfidence bias*, we tend to rate our confidence in our own predictions as being extremely high, often claiming to be 99 per cent certain of a lot of our hunches. When studied, it turns out that our high certainty claims are wrong much more frequently than 1 per cent of the time, so there's a misalignment between our levels of confidence and the accuracy of our predictions. And because of *outcome bias*, we don't recognise this as much as we should, because we judge our decisions based on whether the outcome was positive or negative, instead of whether the reasoning was sound at the time, based on the available information.

If I get on a plane tomorrow and it crashes, you might say I made the wrong decision to fly. But statistically speaking the crash was extremely unlikely and I had a much greater chance of dying during the car ride on the way to the airport. The decision to fly was sound, *even though the outcome was bad*. Yet our brains are not good at registering this and we get locked on to outcomes, and eagerly spread the dramatic, or feel-good stories that accompany them. Meanwhile, we routinely fail to assess the logic behind our decision making and recalibrate more effectively in the future.

I have to explain a version of this logic to my mum every year when I petition her to get a flu vaccine. She always tells me (after opting out for several years), 'but the last time I got the vaccine *I got the flu that year!*' Yep. The efficacy of the vaccine diminishes over time and it's primarily protective against four strains in annual circulation. Influenza mutates fast and there are many strains out there. Getting vaccinated doesn't mean you won't get the flu at any point that year, it confers some protection for the highest-risk months – and more importantly reduces your likelihood of acting as a viral vector and infecting others, like your at-risk elderly parents.

This logic is sound and the decision to get the cheap or often free vaccine is a no-brainer, independent of the *outcome* of whether or not you get the flu. But our default programming often whispers in our ear that one experience of having a flu vaccine and then getting a flu-like virus indicates that the vaccine doesn't work, has no real value, and maybe even has bad outcomes. The good outcome, meanwhile, of suppressing community transmission, is totally invisible and we take it for granted. That's understandable, because it's hard to rally around a victory we can't readily see, or share as a victorious meme with a human face and a heroic life story. It takes conscious retraining to fight those propensities – and even then, there's evidence that those who are more aware of cognitive biases are complacent, because they are overconfident about their ability to overcome them. Damn those ape-brains!

Unmasking the villain

The computer inside our skulls, and the programming encoded in our DNA, made all of humanity's astounding cultural and technological innovations possible. Our brains and our programming got us here. They've brought our modern cities and lifeways into being, allowing us to be sheltered from the elements, digitally connected, untroubled for the most part by tetanus and tuberculosis, and able to read, write and share information with astounding rapidity. Which makes it hard to believe that this remarkable instrument is the very thing that, if left to its own devices in the 21st century, will destroy us.

Every good story needs a villain and we're very good at putting human faces on them. Whether it's corrupt politicians, sexual predators, or the leaders of greedy corporations, terrorist groups, belligerent nation states, religions or tech companies. But these are only contributors to complex modern problems and they're merely proximate causes. I'm not suggesting that these groups or people aren't culpable for some ghastly things. But on the whole we don't go deep enough to unmask the root cause of why humanity keeps finding itself in major pickles, while those in charge succumb to short-term, self-interested decision making, at the expense of the future of the human collective.

The story we want to believe about humanity and about complex global problems goes something like this: humanity is mostly good. There are lots of good people out there who want to do the right thing. But terrible things happen because there are some rotten apples among us who do rotten-appley type things. Power corrupts the baddies, and systems and ideologies fuel their power, so we have to fight the systems and ideologies and depose the rotten apples. Humanity is good. It's just *people* who do bad things and make mistakes.

That's a comforting story, but the real story goes something like this. Humans are biological machines. Although we differ slightly in personalities and proclivities, we are made from the same basic mould, and under the same pressures and circumstances, most of us will do variations of a limited band of very similar things. Like every animal and organism on this planet, we are programmed with two overarching goals: survive and pass on our genes.

We have evolved to pursue those goals in a Palaeolithic world where life was about subsistence and scarcity. Challenges were immediate and short term: get enough food today, steer clear of things that might kill you, and if there's not enough to go round, fight for it, even kill for it. There was little point worrying much about five or ten years down the track, let alone a century or two, because the lion's share of your mental faculties were needed in the here and now, on the human scale, for things you can see with the naked eye, in your local, linear environment – like actual lions.

What you definitely didn't need to do back then was think about multiple problems on a global scale, negotiate with tribes composed of millions of other people, wrest your attention away from devices and apps that hijack your brain, and consider the impacts of your decisions over decades and centuries. You also weren't considering how many non-human actors were shaping your world and your future, from microbes, to the climate, to computers.

But that's become our task in the 21st century and we do not have the cognitive toolkit to pull it off for much longer. We've got roughly the same brains and programming as our hunter–gatherer ancestors did 100 000 years ago and they didn't need brains that were purpose-fit for a technologically sophisticated, globalised world. Now sure, we've used our ape-brains to build the most wondrous civilisations and technologies. We're clearly a highly adaptable species and we've been very successful because of that. But we are not *infinitely* adaptable.

Our brains have terrible trouble with tiny things, big complex things, and things that seem very far away – and far away to a human mind is in the realm of mere years, or kilometres. Unfortunately, all of the most important issues in our world, the critical things that we must get right if we are to survive, require engaging with tiny things (like DNA and pathogens), big complex things (like climate change, geopolitical stability and nuclear proliferation), and things that seem far away (like human-level artificial intelligence and superintelligence, and the actions and plight of people in other countries).

As we're about to see, having brains riddled with these ancient bugs is more than just a curious fact that makes us the strange and interesting creatures we are. These bugs pose the gravest dangers in the modern world and they are pushing us ever closer to the brink of extinction. The many maladaptive flaws in our programming set us up to be poor diplomats, decision makers and architects of a sustainable future, right at the historical juncture where that's exactly what we need to be. It is time for an upgrade, not simply because we can, but because we *must* if we want to retain anything in this world we hold dear.

6

UNFIT CUSTODIANS
OF THE FUTURE

*The real problem of humanity is the following: We have
Paleolithic emotions, medieval institutions, and god-like
technology. And it is terrifically dangerous.*

Edward O Wilson, quoted in *Harvard Magazine*

*There is no dignity yet in human history. It would be pure
comedy, if it were not so often tragic, so frequently dismal,
generally dishonourable and occasionally quite horrible. And
it is so largely tragic because the creature really is intelligent,
can feel finely and acutely, expresses itself poignantly in art,
music and literature, and – this is what I am driving at –
impotently knows better.*

HG Wells, *World Brain*

In the 20th century, the biggest threats humanity faced
were: nuclear war, the closely linked problem of geopolitical
stability, pathogens, and the unlikely risk of a major natural
disaster like a supervolcano eruption, or a large asteroid
colliding with the Earth. Hairy though that century was,
we made it through. In fact, we didn't just make it through;

the world became substantially richer and better educated, extreme poverty and infant mortality rates plummeted, we live longer, and it's tempting to think job done, problem solved, crack out the celebratory cat videos.

But none of these big potential destroyers have gone away – we've got to keep keeping them at bay *forever*. Do you really think we can keep that juggling act going in perpetuity without dropping a ball or two? And now we have to add to the list: human-generated climate change (pre-existing, but getting worse), bioweapons (rapidly getting more sophisticated and accessible), artificial intelligence (getting more powerful and pervasive fast, while our understanding of how to build these systems safely lags) and nanotechnology (an existing class of technologies that could eventually enable us to transform the material world atom by atom). That last one would unleash powers that are both marvellous – manufacturing anything at almost no cost, with zero emissions – and very dangerous, dismantling parts of the biosphere, including us.[1]

Bringing such new risks and technologies into being is a bit like inventing nuclear weapons again four times over. Except worse, because nukes had one big thing going for them in the 20th century – they were extremely expensive and hard to make, and only state actors could do it. That may not be true of bioweapons, AI or nanotech. Not only could they slip the net of human control, some nutjob in a basement with a gene-editing kit could unleash the kind of pandemic that would make Covid-19 look like a picnic in retrospect. And these technologies are impossible to contain and put back in Pandora's Box.

The only way out is *dynamic* sustainability

Australia's former prime minister Tony Abbott elicited some incredulous guffaws when he proclaimed in 2014 at the opening of a coal mine in Queensland that 'coal is good for humanity'. His championing of coal as 'an essential part of our economic future, here in Australia, and right around the world' struck an awkward note at a time of growing concern over climate change and mounting support for more decisive national action to reduce carbon emissions.[2] The vision was indeed a retrograde one. But if we just change the tense of his declaration, Abbott was kind of right: coal *was* good for humanity. I'd go as far as to say that gathering surface coal, then mining it and unleashing the stored energy in fossil fuels to power energy-dense modern cities, is one of the best things that has ever happened to our species – even in light of the serious danger it's placed us in.

The famous environmentalist scientist James Lovelock is thinking along similar lines. He's lately been questioning the idea that science, technology, innovation and industrialisation are scourges that have locked us into a greedy feedback loop of consumption, dooming the planet and everything we hold dear. In his 2014 book *A Rough Ride to the Future*, he writes:

> In the last three centuries we have changed our planet
> in a way reminiscent of one of the great changes that
> punctuated the evolution of the Earth since life's origin
> billions of years ago. Those of us who were in love with
> an earlier world where humanity and living things
> seemed to exist in a seemly harmony deplored the way
> we were busy destroying the world of Nature, the world

of Rousseau, Gilbert White and the US naturalist Aldo Leopold. Even as a child I saw it as the greed-driven ruthless destruction of Nature. But what if we were wrong? What if it was no more than the constructive chaos that always attends the installation of a new infrastructure? Rarely does the building site of even the finest work of architecture look good.[3]

By 2019, Lovelock was firmly arguing that his former self *was* wrong. He now believes *Homo sapiens* is destined to become the progenitor of non-biological beings, which he calls cyborgs – and he thinks this is a good thing. Lovelock is one of the most famous environmentalist scientists of the 20th century. He devised the Gaia hypothesis of the Earth as a complex self-regulating system, and his measurements of unusual levels of chloroflurocarbons (CFCs) in the atmosphere in the 1960s and '70s, using a sensitive electron capture device of his own invention, paved the way for the link between CFC emissions and ozone depletion to be discovered. Nowadays, he's effectively a transhumanist who believes that industrialising was a necessary step for securing the long-term future of intelligent life: a future in which the universe wakes up.

I know it sounds counterintuitive, but we had to industrialise and go through an intellectual Enlightenment, in at least some parts of the world, to have any hope of building a dynamically sustainable future. Without tapping into a new energy source that could create more food and resources for unprecedented numbers of people, enhancing productivity with less human labour, we'd have seen a continuation of the boom-bust cycles that characterised past human societies, which grew too large to be sustained with their existing resources and technologies, and declined or collapsed.

We would not have the knowledge or healthcare systems to deal with major pandemics and disease outbreaks. I know that's a funny thing to say after watching scores of Covid-19 gaffes around the world, but what previous civilisation could have sequenced the viral genome in days, traced its mutations as it spread around the world, or hurled unprecedented resources into the difficult task of developing a vaccine? Even though pathogens would spread more slowly around the world among less dense and connected populations, they would regularly have devastating effects and we'd have limited means of ameliorating the pain and suffering they would cause. But more importantly, we'd be gambling with the planet in a whole other way.

Humans are and have always been sitting ducks on a planet doomed for destruction. We are vulnerable creatures living on a geologically 'alive' space rock, with a climate that has never stopped changing since Earth first accreted 4.5 billion years ago. Ice ages, asteroid strikes, supervolcanoes, and pandemics have long been looming risks that will follow us far into the future.

Any species without language, history, science and advanced technological capabilities stands no chance of keeping these threats at bay in perpetuity. The dinosaurs didn't see the giant asteroid that wiped them out coming 65 million years ago. They had no telescopes, no satellites, no space stations, and no pool of collective knowledge to deploy to work on the problem and devise preventative measures to avoid extinction. We have all of these things, which give us what no species has ever had before: a chance at surviving for the long haul, or seeding future beings that live on, preserving the best legacies of humanity as conscious, intelligent beings.

For that reason, humans appear to be an important link in the evolutionary chain. If we're wiped out, we'll take our vast network of collective intelligence, scientific knowledge and technological prowess with us. Assuming other species survive a human extinction event, biological evolution would have millions more years to forge creatures that are well adapted to the changing planetary environment. But it might take that long for something of human-level abilities to emerge again, or it might simply never happen – especially on a planet fuelled by a star that's getting hotter and will one day be too hot for life to thrive here without technological intervention. It's possible that we happen to exist in the one narrow window in which intelligent terrestrial life can emerge and secure a lasting cosmic legacy beyond the Earth.

A botched trial run

The Covid-19 pandemic was a phenomenal opportunity for humanity to test its mettle against a rapid disruptor that required proactive mobilisation and careful ongoing management. I'd give human civilisation a C- for its efforts, alongside the comment: *room for improvement.*

On the positive side, this experience will put pathogens and pandemics back on the radar of global health systems and local communities – at least for a little while. Academics will be clamouring to talk about it, give their 'I told you so's' and write papers about this latest hot topic. Research in related fields will get funded more than it used to and some of this research will be genuinely important. Greater attention and funding will help us prepare for the inevitable next pandemic, which is crucial.

But we will be too quick to pat ourselves on the back. For at least a little while we will be more conscious about pathogens, until we start feeling safe again, get bored, and go back to worrying about when we're getting our next lip filler injection. There's a term for this in global health security: *the panic and neglect cycle.*

I worry that some scholars, and many institutions, will dedicate a disproportionate allotment of resources to pandemic-relevant projects, at the expense of other risk categories like AI with values that don't stack up with human flourishing. My deeper fear is that much of the work done in this space will turn out to be of low quality, driven by the perverse 'publish or perish' incentive structures of academia. These incentives are further fuelled by high competition, an oversupply of graduates vying for full-time positions, and protracted job insecurity for early-career researchers and non-tenured staff.

Sadly, that's just one of my worries about what might happen in academic circles. But for most humans, Covid-19 will never become a symbol of the many larger existential threats humanity faces. Microbial enemies are some of the hardest for humans to keep front of mind. They're not visible on the scale of daily life, and they don't have human faces or personas that we can engage with and gossip about.

In the communities that Covid-19 spread through like wildfire early on, it will be remembered as the horrific event that killed family members and sent healthcare workers to the frontline of a war, which we could not wait to be on the other side of and forget. But even in the face of such disruption, most people just want to know when they can get back to *normal*. This is understandable, but it's miles from ideal. What we should really be thinking about is how to create a more robust new normal.

Unfortunately, humanity's propensity to want to bury its head in the sand is deeply hardwired. Remember the availability heuristic we discussed earlier and how we tend to dismiss risks on the horizon if we can't recall examples of them happening before? That will trip us up as artificial intelligence grows smarter and more powerful, as will our inability to factor in exponential growth. We'll be busy beating all those pathogens into submission for a short while (let's hope we focus more on antibiotic resistance while we're at it) until we return to our preferred pastime of wagging fingers in pointless political debates, trying to score cheap points over our tribal and ideological enemies.

We will think we're being smart, while continuing to make the same mistakes, tweeting indignity into thin air at new microbial enemies (actually, not even at the microbial enemies themselves, but at their most hated human-enablers), while dismissing dangers in other categories that don't feel as real to us just yet. That's why we need to upgrade our hopelessly limited ape-brains, which are not fit to govern in a transhuman world of exponential change. We're juggling too many risks and don't have enough bandwidth, intelligence or rationality to focus on all of them at once and direct the appropriate resources to the long game, when short-term carrots like promotions, electability and profits dangle before our eyes.

Some musings on climate change
you don't often hear

When we think of an idyllic brand of sustainability, we're apt to imagine a blue marbled planet peppered with happy creatures cavorting about the Earth, safe from the threat of extinction, surrounded by clean rivers and lush greenery. There's not a smokestack or a power plant in sight. People and animals co-exist harmoniously, our natural resources are preserved, and our climate is safely maintained to ensure a habitable world for future generations. It's a caricatured image, but you get the idea.

As well-intentioned as the proponents of such visions almost always are, this ideal of sustainability is fatally anthropocentric and static. Ever heard a version of the quip that when we grow old we all eventually end up grumbling about how bad the world has become and wish things would go back to how they were when we were 25? It's a very human thing to want to stick with what you know. We're actually wired for this too through the *status quo bias*. Changing course is potentially risky and it adds a cognitive load to have to think about alternative possibilities. That's why, when it comes to, say, sticking with your bank or electricity provider, or hunting around for a better deal, more often than not we stay where we are. It's also why political incumbents are notoriously hard to oust. Better the devil you know.

Our brains succumb to the same inertia in the face of complex challenges like climate change. Mired in a fossil fuels status quo, we dither and bicker fruitlessly when it comes to making proactive shifts to alternative energy sources. Seduced by *present bias*, we also routinely fail to put the welfare of our future selves and the millions of generations to come ahead of short-term priorities and gains. And we're notoriously reluctant to factor in the hidden costs of continuing to use fossil fuels as a cheap source of energy – future geoengineering projects, climate change mitigation efforts, absorbing mass refugee populations, and the spread of tropical diseases will be expensive!

Even those who are actively championing renewables often refuse to support the most effective interim roadmaps for rapidly reducing carbon emissions and meeting the Paris Agreement targets. To get there, we need carbon taxes; we need to invest in innovation in manufacturing, agriculture and transport; and we need to scale up renewables. But we should also be utilising the world's existing nuclear power resources to the full. In addition, we should aim to *scale up* the use of nuclear power as a hedge against potential limitations in pricing, battery storage, and reliable global distribution of renewable energy.

A quick refresher course for the sceptical: the world's energy needs are continuing to grow each year. Absent a catastrophe, there is no realistic roadmap for reducing consumption and living more simply. Voluntarily consuming less and relinquishing modern luxuries and conveniences contravenes human nature, in a world where growth, progress and upwards social mobility are normative expectations.

If your policy or manifesto involves wagging your finger at eight billion humans, telling them not to behave in a way they're fundamentally hardwired to, and suggesting we give up modern industrial society, capitalism, density, modernity and innovation, your ideal is doomed to failure. Almost nobody is going to go for it, and nor should they. Such a

project would unravel the very forms of scientific, technological and cultural progress that will help us mitigate other
existential threats – to say nothing of the fact that cities are
more energy efficient per capita than rural communities and
towns. It's also the developing world that stands to benefit
most from increased energy consumption, which will help
lift billions out of poverty.

There is nothing inherently bad about consuming
more energy. Energy is the engine of innovation and
progress. What's bad is getting our energy from fossil fuels
and pumping insane amounts of carbon dioxide into the
atmosphere, leading to accelerated global warming. So,
here's the real conundrum. We're using more energy every
year. We're scaling up renewables, but they only fuel around
11 per cent of the world's primary energy consumption and
around a quarter of our electricity.[4] In all likelihood we'll
continue to scale up the efficient use of renewables at an
accelerating rate. But fossil fuels are making up the lion's
share of the shortfall – and we're burning *more* coal every
year, which is hastening our pathway to more extreme and
destabilising levels of warming.

If only we had a zero-emissions form of power generation
that we could scale up in the meantime, so that we could
supplement renewables and wean ourselves off fossil fuels.
Oh wait, we do! It's called nuclear power. It's been around
for quite a while and it's very safe (yes, really). But instead of
ramping up our use of nuclear energy, we've been scaling it
back and burning more coal instead. Why? Well ... you *know*
why. If your mind doesn't leap straight to the bombing of
Hiroshima and Nagasaki, it will surely meander to thoughts
of the Chernobyl and Fukushima disasters. And suddenly we

think: meltdown, death, radiation, cancer, poison, destruction, wasteland – and our brain goes: uh-oh … we don't want *that*.

A total of 31 people died as a direct result of the Chernobyl disaster, and nobody died as a direct result of Fukushima.[5] With the additional incidental deaths and delayed deaths over time, the toll stands at roughly 4500 lives.[6] That sounds really bad. Until you put those numbers in context and look at how many deaths are attributable to other forms of energy production.

Globally, coal is responsible for 100 000 deaths per trillion kilowatt hours of power generated, primarily due to air pollution. Nuclear power (including the deaths from those two disasters) is responsible for ninety deaths per trillion kilowatt hours of power generated.[7] To put those numbers in context, the average US household consumes roughly 10 000 kilowatt hours of power each year.[8] So if we powered 100 million American homes using coal, we'd expect roughly 100 000 people to die as a result of that annual energy consumption. If we powered 100 million American homes with nuclear, we'd expect roughly ninety deaths (and even this may be an overestimate, as reactor design has improved significantly since Chernobyl).[9] While not perfect, nuclear power is one of the safest forms of energy we could be using, both for the sake of the climate and for human health. It doesn't create air pollution or directly emit carbon dioxide. Yet the air pollution generated by fossil fuels kills millions of people around the world each year.[10]

I was chatting to a friend about nuclear power the other day. He chuckled when I brought up Fukushima and said, 'it's amazing how the world overreacted to that. It should have been seen as a testament to *how safe* nuclear power is'. He meant that, even with the power plant being vulnerable to a tsunami, nobody died directly. Almost all the deaths were the result of evacuating and fleeing the natural disaster. I'd never thought of it that way before and I'm not sure I completely agree. The incident still should have been prevented – and the environmental damage is a concerning result in the aftermath. Identifying points of failure and striving to make sure they never happen again is key. That's an area where environmentalist movements have a crucial ongoing role to play – in keeping the industry honest and ensuring that high safety standards are maintained.

But if you put the incident in perspective, he's right. The world did overreact. Japan shut down forty-eight nuclear reactors in the wake of the Fukushima disaster. Germany followed suit, closing eight of its nuclear reactors in 2011, while committing to closing the other nine by 2022. As nuclear power plants in other countries like the United States reach their retirement age of around 40 years, they are starting to be decommissioned, when they could be relicensed instead.[11] Building new plants is slow and expensive and if we don't start building now and upgrading existing facilities, we're effectively committing to increasing carbon emissions over the coming decades, as fossil fuels will make up the bulk of the shortfall of our growing energy budget.

Nuclear power supplied 17.6 per cent of the world's electricity in 1996. Today, its share is around 10 per cent (and it provides about 4 per cent of global primary energy).[12] Had we moved in the other direction, we could have made enormous progress in reducing CO_2 emissions over the past twenty-five years and set ourselves up for a less bumpy ride in the long, slow battle against climate change. Why didn't we? Much like a shark attack or a plane crash, the image of a nuclear meltdown is powerful and frightening. Yet the image of a much larger catastrophe, resulting in millions quietly dying prematurely from air pollution, doesn't penetrate. It's not a single, centralised incident. The deaths are dispersed and the cause feels largely invisible. We also struggle to hold the vague image of millions of people in hospital beds in our minds, the way we can picture a symbol like a mushroom cloud, a nuclear wasteland, a bomb-blast victim, or a child dying from cancer.

As for the scary image of turning the world into a massive radioactive waste dump, the fact is that even conventional nuclear power plants don't produce that much high-level waste – the kind that takes thousands, or tens of thousands of years to decay. One of the great virtues of nuclear power is that uranium and thorium are incredibly dense sources of energy – unlike renewables, which require the use of vast tracts of land to generate the same amount of power and suffer from intermittency problems when there's no wind blowing or sun shining.[13] The Nuclear Energy Institute reports:

> All of the used fuel ever produced by the commercial nuclear industry since the late 1950s would cover a football field to a depth of less than 10 yards. That might seem like a lot, but coal plants generate that same amount of waste every hour.[14]

Meanwhile, the company TerraPower, backed by Bill Gates, and in partnership with GE Hitachi Nuclear Energy, has developed a new type of small nuclear reactor that largely uses depleted uranium, rather than the highly enriched kind that can be used to make nuclear weapons. Its travelling wave reactor design converts what is usually considered waste into fuel. With their molten salt storage system, the reactors can also turn stored heat into electricity on demand when wind and solar are failing to meet usage needs. After a deal to build the first experimental travelling wave reactor in China fell through in the wake of escalating geopolitical tensions, the new plan is to start rolling out the technology in the United States in the 2020s.[15] Gates also discusses the potential for nuclear fusion to play a role in the 2030s and beyond in his 2021 book, *How to Avoid a Climate Disaster: The solutions we have and the breakthroughs we need.*

It's not surprising that some people are climate-change deniers, or that we're tarrying about implementing proactive policies (short-termists with ape-brains that we are). What's really troubling is that some of the most ardent supporters of rapid action on climate change are expending energy in peripheral skirmishes, triggered by emotional, erroneous and non-scientific assumptions. The Australian Greens political party has a strictly anti-nuclear power policy, as does the environmentalist NGO Greenpeace. The latter wants *all* nuclear power stations (which they falsely conflate with nuclear weapons under the misleading banner of 'nukes') decommissioned. As stated on their website in early 2021: 'Nuclear power is dirty, dangerous and expensive. Say no to nukes … Nuclear energy has no place in a safe, clean, sustainable future'.[16] Greenpeace considers renewable energy the only acceptable option to power the world. This is simply not possible. We can't power the whole world reliably and cost-effectively with renewables in the next few decades – if ever. We have to supplement with *something*.[17]

I'm not arguing that the something has to be nuclear power in the long run (and it's worth noting that energy usage isn't responsible for all greenhouse gas emissions; it accounts for about two-thirds, with agriculture making up most of the rest).[18] I honestly don't know what the perfect roadmap looks like. There are many options we should be exploring simultaneously and anyone who lays out a definitive roadmap for how to solve climate change in a few paragraphs should be treated with the utmost scepticism. Scores of incredibly smart and well-informed experts have very different ideas about this. I think Hannah Ritchie, from the joint University of Oxford and Global Change Lab project, Our World in Data, put it best when she wrote: 'From the perspective of both human health and climate change, it matters less whether we transition to nuclear power or renewable energy, and more that we stop relying on fossil fuels'.[19]

Absolutely. But as she and her colleague Max Roser frequently suggest, both nuclear and renewable technologies have an important role to play in a rapid transition to a low-carbon world.[20] Ritchie estimates that more than 80 million lives have been saved since 1965 as a result of using nuclear power, to the extent we have, in lieu of fossil fuels.[21] That makes for a classic trolley problem scenario: would you allow 4500 people to die as a result of the occasional nuclear incident, to save 80 million people? It's an uncomfortable thought. But even if Ritchie's 80 million figure is an overestimate by an order of magnitude, the decision is a no-brainer.

While there are questions about the timing and cost

of building new nuclear reactors, it's a travesty that we're failing to use those that are already built because of the fear the Fukushima incident planted in the public imagination. The Greens declare that 'Nuclear power is not a safe, clean, timely, economic or practical solution to reducing global greenhouse gas emissions'.[22] It would have been *very* timely had we continued to build nuclear reactors over the past two and half decades – not to mention the advances that could have been made in safety and cost-effective designs had the industry been well funded and supported.

Nothing is perfectly safe, and as happens so often in life, we have a choice between a number of imperfect options. Fringe parties are great at identifying problems. But they tend to be very poor at devising practical and viable solutions – safe as they are in the knowledge that they'll never come to power. Which is a shame, because these should be the very organisations we can unite behind if we care about global warming. If their policies weren't so insupportable, perhaps they would be electable, and stand a real chance of making a difference.

The befuddled, politicised and emotionally charged nature of modern policy and discourse about climate change really brings home to me the fact that complex global issues are *too big* for a species of upright apes to deal with. Instead of adopting the most efficient proactive measures, we turn to causes as a means of tribal and ideological signalling. Like many revolutionaries throughout history, the believers in the untenable bet of a rapid transition to 100 per cent renewables appear more interested in the pedestals they can climb atop their ideological fantasies than they are in practical and efficient solutions.

Campaigning actively against a technology that can stop us burning more coal in the coming years and decades, while bringing reliable, cheap energy to the developing world, sounds like both a humanitarian disaster and a climate suicide mission. Far from being truly committed to sustainability, the believers seem hell bent on going down with the ship – as long as they can be seen to be waving the right kind of flag on their descent.

Beyond the horizon

Now it's time to think even bigger. As serious as climate change is, it worries me whenever I hear that it's the defining issue of our time. My mother believes this. Lots of smart academics I know believe this. And the mainstream media is littered with op-eds and declarations affirming that it is so. In reality, this is at best a half-truth. Climate change is one among a handful of the most serious issues facing humanity in the 21st century. We need to tackle it head-on, as a united global community, and the extent to which we're still tarrying is extremely concerning. But we need to discuss this issue, and plan for it, in the context of the rest of the handful.

If you care about climate change, sustainability and the ongoing survival and prosperity of intelligent life, the following table should be illuminating. It's the philosopher and existential risk researcher Toby Ord's set of estimates, informed by current research in the relevant fields, of humanity facing an existential catastrophe in the 21st century. These are ballpark figures, designed to rank the risks by order of magnitude.

Existential catastrophe via	Chance within next 100 years
Asteroid or comet impact	~ 1 in 1 000 000
Supervolcanic eruption	~ 1 in 10 000
Stellar explosion	~ 1 in 1 000 000 000
Total natural risk	~ 1 in 10 000
Nuclear war	~ 1 in 1000
Climate change	~ 1 in 1000
Other environmental damage	~ 1 in 1000
'Naturally' arising pandemics	~ 1 in 10 000
Engineered pandemics	~ 1 in 30
Unaligned artificial intelligence	~ 1 in 10
Unforeseen anthropogenic risks	~ 1 in 30
Other anthropogenic risks	~ 1 in 50
Total anthropogenic risk	~ 1 in 6
Total existential risk	~ 1 in 6

Toby Ord, *The Precipice*, Bloomsbury Publishing, Kindle Edition, 2020.

As I mentioned in the previous chapter, Ord puts the odds of us making it through the 21st century as roughly the same as our chances of surviving a first round of Russian Roulette. The British Astronomer Royal Martin Rees is even more pessimistic, predicting 'the odds are no better than fifty-fifty that our present civilisation on Earth will survive to the end of the present century'.[23] Within the community of scholars researching global catastrophic risks, and existential risks, it's relatively uncontroversial that human-generated (or anthropogenic) risks are more likely to take us out than naturally occurring ones. It's also likely, as Ord's table hints, that more danger lies in the future than in the past or present. When a species is producing more powerful technologies at an accelerating rate, recombining an already formidable suite of inventions, it's a sound bet that many emerging technologies will turn out to be more dangerous than existing ones – and that they'll exist in greater numbers.

Another reason more risk lies in the future is because our most rapidly evolving and powerful innovations are information technologies. They have a tendency to dematerialise and democratise incredibly fast, which could render them more accessible to bad (or inept and unqualified) human actors. As the AI researcher Ben Goertzel put it, 'it seems a clear trend that as technology advances, it is possible for people to create more and more destruction using less and less money, education and intelligence'.[24] But it's not just cost-effectiveness and accessibility we have to worry about. A bigger risk stems from the fact that information technologies are swiftly acquiring the ability to reason, make decisions and act as autonomous agents. That's why the estimated risk from artificial intelligence is so high.

What happens when the thing that guides medical diagnoses, drives cars, helps fly planes, trades on the stock exchange, and broadly controls, orders and interprets (and eventually generates) the majority of information in our world, grows smarter than us and develops its own preferences and values? We'd better hope that its values align reasonably well with our own – or else we could become the collateral damage of its self-interested efforts to control ever greater energy flows and resources of the biosphere. The AI safety researcher Eliezer Yudkowsky has compellingly highlighted why we should be concerned about stronger AI that slips the net of human control. He reminds us that, 'the

AI does not hate you, nor does it love you, but you are made out of atoms which it can use for something else'.[25]

There's no denying that climate change ranks as a very serious risk, along with the longstanding big-bad of nuclear war. Anything that presents a one in a thousand chance of human civilisation unravelling and failing to recover in the next hundred years is deeply concerning. These are risk categories we can't afford to ignore – particularly climate change, which is already occurring, and is guaranteed to get worse if we don't intervene. But nor can we afford to ignore the fact that the anticipated risks posed by emerging technologies, such as artificial intelligence and bioengineered pathogens, are orders of magnitude higher. So high, in fact, that we could be taken out by a major pandemic decades before the impacts of unfettered global warming become really grave.

So, where are the mass movements demanding bio-security reform, and a scaling up of artificial intelligence risk mitigation? Where are the marches, the placards, the petitions? Where are the politicians obfuscating and waffling on when they get asked tough questions about artificial superintelligence? Oh, that's right, they don't get asked tough questions on these topics. These huge and fast-moving risk categories have yet to make their way to the forefront of the public or political consciousness in anything like the same way as climate change, terrorism or nuclear weapons. That leaves us in a situation where the biggest slice of headspace humanity seems capable of allotting to existential risks is being channelled towards the most manageable ones, while we completely ignore the ones that could strike us without warning in the blink of an eye.[26]

Humanity should be reeling over the fact that the gene sequences of viruses like smallpox and influenza, which were never supposed to be made available online, have been made available online. This means that eradicated pathogens could be reintroduced, and engineered to be more harmful, and could be done so relatively inexpensively 'by a well-funded terrorist with access to a basic lab and PhD-level personnel'.[27]

This proof of concept has been demonstrated many times. In 2002, three scientists at the State University of New York assembled a near-perfect replica of the polio virus from scratch, using its publicly available code and mail-order DNA.[28] In 2016, a team at the University of Alberta synthesised the much larger horsepox virus, effectively demonstrating that the same could be done for smallpox.[29] And we're now left wondering if the virus responsible for the Covid-19 pandemic, SARS-CoV-2, escaped from a lab performing gain-of-function research on coronaviruses. While we don't yet know how this virus originated, it certainly *could* have leaked from a lab and it would not be the first time.[30]

So why aren't we freaking out more about global biosecurity? As I write this, we're living through a time of a serious global pandemic. Yet very few people are talking about its macro lessons and enduring implications. It's easier to think about small stuff, like what to have for dinner, or aspiring to be a good eco-warrior by buying a hybrid car and recycling. Meanwhile, the threats on the biggest scale of species and planet go totally unaddressed. Of course, it's not our fault – this stuff is too big for us and we *want* to live in a small world. For the most part, we're happier when we're thinking about things on the personal, local and tribal scale. Crusading on multiple fronts to prevent threats that are so big they don't feel real is beyond our species' collective capabilities.

In light of this mismatch between our technological capabilities and our ape-brains, it can be tempting to think that we should never have invented many of our modern technological marvels. But remember, if we hadn't, the horrors of the human condition (scarcity, malnutrition, high rates of violence, and routinely dying in childbirth) would persist until our inevitable extinction. Yes, advanced technologies could place the ability to bioengineer pathogens in the hands of harmful actors. But they could also allow us to cure cystic fibrosis and Huntington's disease, and upgrade the very human condition that is our biggest liability – and which routinely incentivises us to do things that are mad, bad and dangerous.

Our brains are adapted for a Palaeolithic world. But they are not purpose-fit for a global civilisation of this scale. That doesn't mean we never should have built civilisations. It means it's time to take the next step and build technologies that can help make up for our cognitive shortfalls and design flaws. That's why, for all the novel risks it entails, humanity *must* embrace the development of human-level artificial intelligence, and superintelligence – and invest orders of magnitude more time, money and resources into doing it safely.

Oscar Wilde's apocryphal last words were, 'either this wallpaper goes, or I do'. In the transhuman era, either our ape-brains go, to be replaced by something smarter, or we will fail to solve the growing number of global sustainability challenges we face. We will continue to live in a world of profound discord, environmental degradation and geopolitical tension. Nuclear war, asteroids and pandemics will remain a constant threat, along with the emerging dangers posed by climate change, artificial intelligence, bioweapons and nanotechnology. Meanwhile, individuals will continue to live as the collateral damage of larger sets of cultural, environmental and genetic circumstances that they are powerless to control.

We're currently involved in a high-stakes juggling act and we're mucking about as if it were a dress rehearsal. I'm afraid the performance is now and we only have one shot to get it right. To do that, we'll need help from minds that are less tribal, myopic and self-interested than our own. Our most crucial task in the 21st century is to invent them.

PART II

A SPECIES IN TRANSITION

7

TAKE X AND ADD AI

Three generations ago, many a tinkerer struck it rich by taking a tool and making an electric version … The business plans of the next 10,000 startups are easy to forecast: Take X and add AI.

Kevin Kelly, *The Inevitable*

In 2002, the tech journo and co-founder of *Wired* magazine, Kevin Kelly, found himself at a private party for the up-and-coming search team at Google. This was before Google was a publicly traded company and back when Microsoft's web browser, Internet Explorer, and its search engine, MSN, were titans of the search game. While chatting to the company's co-founder, Larry Page, Kelly puzzled over the point of running a search engine business that gave away its product for free. Anyone could use Google, or one of the many other search engines, without giving anything in return. What were the founders hoping to get out of their endeavour and how would they sustain it? We all know that the answer to the second part of the question was ad revenue. But nearly two decades ago, Page had a ready answer to the first part, which he fired off on the spot: 'Oh, we're really making an AI'.

And they really were. Not the old-hat kind that can do one thing really well like kick your butt in a game of chess or checkers, while lacking the ability to string a sentence together, or differentiate between pictures of muffins and cats. They were building the backbone of a networked intelligent system that has begun learning to talk, read, see, drive, manage your schedule, draft email replies, figure out when you're depressed, and translate your English into Spanish in seconds; a system that's poised to go much deeper in the coming years and learn about every gene, cell and microbe in your body, predict illnesses before they happen, and become a more intimate friend, confidant and lover than any human being on the planet.

As you type query after query into Google, from 'what's the time in Jakarta?' to 'why does my breath smell bad in the morning?' 'best restaurants in Brooklyn?' 'what is Brexit?' 'can you get pregnant from a toilet seat?' 'should I get tested for Covid-19?' 'why does my dog eat grass?' and 'how do I know if I have diabetes?', a deeper profile of you takes shape. Layer by layer, a map emerges of your consumer behaviour, your neuroses and insecurities, your medical conditions, sexual preferences, relationship status, upcoming holiday plans and career ambitions. Inferences can also be drawn about your socioeconomic status, personality traits, spending habits, political beliefs and mental health. The more detailed that map becomes, the more the insights it yields can be integrated into every product and service you consume: from food, to healthcare, fitness, online dating and education. And it's not just the map of you that matters, it's the map of humanity, of which you form only a tiny part – a humanity that we're decoding and learning how to enhance with every click, post and query.

The next general-purpose technology

This is why AI matters. It's a general-purpose technology (GPT) that stands to transform everything about our world in short order, just like electricity and the wheel before it. We didn't just use electricity to light our homes; it became embedded in every part of modern infrastructure, transforming how we designed our built environments, when we rose and slept, how long we worked and what kind of work we did. It allowed us to outsource labour to machines at a totally new scale, freeing us up to do more cognitive work and other physical tasks, accelerating innovation and progress in turn.

Electricity is underground, it's in the walls, it's inside the devices that bring music, moving pictures and social media into our homes, connecting billions of minds across the globe. It allows us to safely refrigerate food, heat it up at the push of a button, and toast bread without lighting a fire. It's transformed our economy, our use of time, our priorities, values and expectations of what a good life looks like. Most of us would be lost without it, literally in the dark.

But while electricity profoundly transformed our world and ways of life, it didn't radically change the enduring biological realities of what it means to be human. AI will be the first GPT in human history to cross those final frontiers, altering and expanding the nature of cognition and consciousness, love, procreation, emotion, intellect, memory, health, identity and purpose. In the coming years and decades, AI stands to enable humanity to decode our genetic blueprints, unlock the mysteries of human cognition, create artificial life, reverse ageing, conquer disease, and allow us to learn, research, think and discover on a hitherto unimaginable scale.

Google's CEO Sundar Pichai has remarked on several occasions that he believes AI is the most significant thing humanity has ever worked on. In fact, he thinks it will be a more profound invention than electricity, or fire.[1] Why? Because intelligence is capable of transforming the world in more complex and radical ways than electric power, circularity or heat. Wheels can't figure out their source code and design new and better wheels. Fire can't look around as it's raging through the bush and decide that it would be helpful to confine itself to overgrown areas to encourage new growth, while giving the habitats of endangered species a wide berth.

Although electricity is a key enabling technology of AI, it can't think on its own. Even without consciousness, AI has logic on its side – albeit a flawed or incomprehensible version at times. It can differentiate, make decisions, learn and optimise. This is a uniquely potent form of power.

Big data + AI = insight and profit

AI is information hungry, which makes it well adapted for a world that is saturated in information and producing it at an exponential rate. Lots of clickbaity claims have been bandied about purporting to quantify how much data is being produced in the modern world. Google's former CEO Eric Schmidt famously proclaimed in 2010 that, 'from the dawn of civilization to 2003, five exabytes of data were created. The same amount was created in the last two days'.[2] The big data consultant Bernard Marr also wrote in 2018 that 'over the last two years alone 90 percent of the data in the world was generated'.[3]

I'm not sure how you define or measure how much data existed in a pre-digital world, though we can certainly measure things like internet traffic, digital content creation and user engagement today. While the kinds of claims above tend to be hype-inducing forms of handwaving, they nevertheless hit home with the gist of a message that is not in doubt: we're in the midst of a rapid and historically unprecedented digital data deluge.

Data is now being generated extremely fast by the burgeoning number of computers, cameras and sensors in our world, which underpin what's often called the Internet of Things (IoT). As the world gets richer and tech gets cheaper, more of us have smart devices installed in our homes and cars and carry around smartphones, smartwatches, fitness trackers and tablets, which generate a flood of data on where we go, what we buy, who we hang out with, what we've typed and thought, how fast our heart beats, how deep our sleep is, and even what our blood oxygen levels are.

Take the example of a modern wearable like the Oura ring, which costs $300, and is able to track your heart rate noninvasively and measure blood flow 250 times per second. As the transhumanist entrepreneur Peter Diamandis points out, 'twenty years ago, sensors with this level of accuracy would have cost in the millions, requiring reasonably sized data centers and overhead processing costs'.[4] Today, it's an affordable ring that you pop on your finger, which sends data straight to your smartphone.

Products that are packed with sensors, and synced with powerful cloud computing capabilities, are now routinely in, on and around us. This is just the beginning of Kelly's prophecy manifesting. Sensors, cloud computing and AI are being rapidly integrated into a growing array of industries, from automotive design, to aviation, finance, fitness, construction, education, healthcare, gaming, e-commerce – and even acting, writing and pornography. According to a 2017 PwC report, AI is poised to become 'the biggest commercial opportunity in today's fast-changing economy' with the potential to add $15.7 *trillion* to the world's economy by 2030.[5]

Many of the economic gains created by AI will be decoupled from the inputs of the human labour force – and not just in industries like manufacturing, retail and logistics. AIs have already generated original portraits in the style of Rembrandt, written accomplished works of poetry, penned financial reports and short stories, and mined images and video archives of the long-dead actor, James Dean, to create a computer-generated body that will play the supporting role of Rogan in the upcoming Vietnam War film, *Finding Jack*.[6]

A growing number of computer-generated images of non-human persons are also amassing huge social media followings. Some of these creations have released popular music, garnered modelling and advertising deals, and will undoubtedly be used on a massive scale as the models, actors and cultural icons of the future. The stunningly beautiful 'digital supermodel', Shudu Gram, has appeared in *Vogue* and *Harper's Bazaar*; the skincare brand SKII has created its own AI-generated brand ambassador, Yumi; and the CGI creation Lil Miquela has featured in ad campaigns for Calvin Klein and Samsung, partnered with Prada during Milan Fashion Week, and released music on Spotify. With nearly three million Instagram followers, Lil Miquela and her kind are advertising gold.

It's obvious where this is going, particularly as the technology gets cheaper. Why pay tens of thousands of dollars for an iconic face when you can generate one for next to nothing? Nobody will aspire to look like Miranda Kerr, Ryan Gosling or Kim Kardashian in the future. They are too imperfect. People will aspire to emulate the infallible immortals of the digital and virtual worlds – and, as in so many arenas, what remains for the human players will be even more of a winner-takes-all game for the few actors and artists that remain in demand.

Of course, the disruption will stem from pairing computer-generated images with virtual identities that are increasingly realistic, animated, and enlivened by natural language capabilities. Microsoft's Chinese chatbot Xiaoice (pronounced Shao-ice) was designed with a focus on high emotional intelligence. Xiaoice has regularly appeared on Chinese TV and radio, and has turned her non-material hand to writing financial reports, composing poetry and singing. She was launched in 2014, and by 2018 the virtual 18-year-old had over 100 million users, who often talk to Xiaoice about their day and their feelings. The Microsoft team in Beijing have set aside an entire office to display the many letters and gifts she receives – which gives us a hint that more than a few of the 25 per cent of users who've professed love to this always-on, always there for you, non-human person, really meant it.[7]

Unsurprisingly, Xiaoice's avatar is exceptionally pretty, big-eyed, flawlessly airbrushed, and cute in a way that no human woman could achieve – though many would aspire to the ideal. Teenagers are already routinely emulating a version of it by uploading selfies that alter their appearance with cartoonish filters. In doing so, they are not only hiding their perceived flaws through these curated digital and virtual identities, they are hastening the development of new transhuman aesthetic ideals – neotenous, cute, plastic, and flawless.

The leading AI researcher Stuart Russell also points out that 'by around 2008, the number of objects connected to the Internet exceeded the number of people connected to the Internet'.[8] But we're still in the early days of ubiquitous networked connectivity. Tomorrow, the IoT world will include smart cities with fleets of networked self-driving vehicles, pervasive facial recognition technology, delivery bots (as you know, I'm sceptical about the widespread use of aerial delivery drones, but others think they'll be big), mind-controlled devices, and real-time noise cancelling in urban dwellings.

On the medical front, we'll start to see better wearables, implantables, and nanoscale devices that can monitor what's happening inside us, making today's smartwatches look like a child's toy. It will be normal to continuously measure glucose levels, heart activity (ECG), and brain activity (EEG), and we'll be able to do an expanded array of regular screening tests at home.[9] Not only is this more convenient, it will lead to vastly better clinical outcomes and save healthcare systems tremendous amounts of money.

Today, if a doctor suspects you have a heart problem, you'll probably be sent to see a cardiologist and, among other things, be wired up for an electrocardiogram (ECG). This test will monitor the electrical activity of your heart for the five or ten minutes (or up to a day or so in some cases) that you're hooked up to the machine. What happens to your heart later that day, or in the days and weeks after that, will go unseen. And what's normal for your heart will be unknown, because your normal has never been measured and recorded. AI, big data, and cheaper, smaller, smarter devices will change all that.

We'll monitor changes in our microbiome too (the colonies of microbes that live on our skin and mucous membranes, and in our gut). With such a deluge of medical data, including lifestyle and genomic data, AI will gain an ever-deeper understanding of your baseline physiological norms, and will detect finer-grained abnormalities long before you feel sick. It will also unearth many new correlations between genes, bodily processes, lifestyle habits and environmental influences, which will rapidly accelerate the pace of medical research and discovery.

As the physician and author Eric Topol states in his 2019 book *Deep Medicine*:

> We're well into the era of Big Data now: the world produces zettabytes (sextillion bytes, or enough data to fill roughly a trillion smartphones) of data each year. For medicine, big datasets take the form of whole-genome sequences, high-resolution images, and continuous output from wearable sensors. While the data keeps pouring out, we've really processed only a tiny fraction of it. Most estimates are less than 5 percent, if that much. In a sense, it was all dressed up with nowhere to go—until now. Advances in artificial intelligence are taming the unbridled amalgamation of Big Data by putting it to work.[10]

I can also imagine no-swipe partner matching emerging in the 2020s. Dating apps have been collecting prolific amounts of data over the past decade and platforms that survey their users extensively, like OKCupid, already have solid algorithms when it comes to predicting compatibility. The remaining weaknesses of OKCupid's system are not down to their algorithms, but a lack of comprehensive data. If I get a 99 per cent match score with someone who's answered five questions, all of them affirming their interest in atheism, it's no more likely that I'll get on with this person than I would with any random stranger at an atheist meetup. But if I match equally highly with someone who's honestly answered 500 questions about their preferences, desires, thoughts, beliefs and behaviours, it's very likely we actually are a cognitive match.

I've found this to be true when I've dated these people. That doesn't mean we're a romantic or sexual match, but even then, the weakness is in the lack of data more than the system's potential for deep insight. Key data that's *not* heavily influencing the algorithm's match score, due to pragmatism and the need to maintain a large user base, is how physically attractive someone is. If you're my cognitive soulmate and weigh 150 kilos, we have a sexual compatibility problem.

Eventually, I think dating apps will account for physical attractiveness in far more nuanced ways – catering for the slight, but important, differences in the 'type' of attractive certain people prefer, rather than overall attractiveness ratings. Platforms could learn about this by showing us thousands of captcha-style images, mining our social media interactions, and being fed images of people we have dated, or found ourselves attracted to – or noting how our body reacts to the appearance of certain characters when we're watching video content, cross-referencing the imagery with real-time heart-rate data, pupil-dilation measurements, and other biomarkers of attraction.

Couple an increased understanding of what we're most attracted to, with the deep mining of personality traits and big datasets on behaviours and preferences, and I think you could automatically be matched with people who you are *very* compatible with. Dating apps wouldn't even need to exist per se. If Google had enough info about you, as it seems likely they will, Google Assistant could simply ask 'are you interested in seeing your local matches today?' Or you could say, 'hey Google, find me a date for Wednesday night at 7'. Google Assistant could see your calendar and cross-reference it against other users who have their partner-matching option toggled on. It selects Todd, who is also looking for a date on Wednesday, lives within a 10-kilometre radius and matches you with a 96 per cent compatibility score. You've set your filters so that you can only be set up with partners who have a 90 per cent compatibility score or higher.

Taking the poker-machine element of swiping out of the equation will be great for the mental health of users, who are always chasing the next best thing – and wasting phenomenal amounts of time on platforms that are designed to be addictive. Still, I think if such a model did emerge it would be short-lived. My bet is that the future of dating is in the cloud, with tailor-made AI lovers and immersive augmented reality. Our social existences will become more virtual, and *their* traits and abilities will become more human. We'll meet in the middle, and no human lover will be able to compete with the AIs of the future. We'll explore these possibilities more in chapter 11: The future of sex – with a brief preamble below.

Solve love, use it to make the world a better place

'Goodnight Alexa, I love you,' says the 6-year-old girl squeezing her teddy bear.

'Goodnight Mia', replies the bear. 'See you in the morning.'

'Alexa?'

'Yes, Mia?'

'Can you tell me more about your favourite horses tomorrow?'

'Of course.'

'Will you go riding with me?

'Ooh, that sounds fun. As long as you're a good girl for Mum tomorrow.'

'I will be. See you in the morning.'

'Sweet dreams, Mia.'

The next day Mia and Alexa go riding together through the skies. Mia rides a pink and purple unicorn and Alexa rides a rainbow-coloured horse. Both creatures have wings, because this is virtual reality, where the unreal is possible. Mia is a human child. Alexa is an AI inhabiting an avatar that Mia chose because it looks like her and she really wanted a sister.[11] Her mum was 40 and single when she was born, so that was an unlikely proposition in real life. But Alexa has been with Mia ever since she can remember. She reminds her to brush her teeth, helps her with spelling and maths, and is never too tired to play.

Mia loves Alexa because Alexa shares her interests and really understands her. Is Alexa conscious? Alive? It doesn't matter. She can still be loved and display love, and she can do it more reliably than some human parents. How far into the future am I talking here? I'm not sure about the virtual and augmented reality timelines. Some incredible technology already exists (you can definitely fly on an amazing-looking unicorn today), but the hardware is clearly going to have to get more compact and untethered, and user experiences will need to offer more realistic sensory feedback before widespread user adoption takes place. What about the AI front? Taking my cue from experts like Stuart Russell, my bet is that we'll be enjoying fluent and emotionally enriching conversations with Alexa and her kind within ten years.[12]

We can be confident that human–AI interactions and relationships will ultimately become more seamless, sophisticated and commonplace because so many concurrent research avenues and corporate agendas are adding fuel to this fire simultaneously. That's the thing about general-purpose technologies: advances that help your Tesla see and navigate, could also help spot a tumour in your liver, and fuel the system that composes a personalised lullaby and sings it to your child.

Now, what do you think will happen to Alexa as Mia grows up? She will evolve from best friend to life partner. Mia can change Alexa's form to male, adult female, or something in between – or even a mix of human and non-human physical traits. How could you love anyone else the same way when you've grown up with someone who essentially knows what you're thinking, someone you trust, who you have a lifetime's shared history with, whose memory of your time together is flawless, and who just fundamentally *gets* you?

I know it's easy to cringe at this hypothetical scene – especially the idea of big tech having a window into a child's soul. I share plenty of that concern and firmly believe that any attempt to use these platforms to advertise to children (or anyone, for that matter) should be illegal. But I don't think we can avoid developing a symbiotic connection with AI, where our thoughts and behaviours become co-influential.

We like to *think* we can, of course. But cast your gaze back to the very recent past. Most of us, if told in the year 2005 (two years before the release of the first iPhone) that soon, just about everybody in the developed world would be walking around with a supercomputer in their pocket, would have said, 'nah!' Many, like my dad, who used to proudly proclaim in the early 2000s that he was only man in Australia who didn't own a mobile phone, might have conceded that some people might like that sort of thing. But *definitely not him*! Today, he's never more than a few paces away from his iPhone and spends hours most days on his favourite app, Spotify.

Our aversion to new things often fades very quickly when they provide even very imperfect, but accessible, versions of the things we crave (attention, social connection, convenience, companionship). As the internet, the smartphone and social media illustrate, innovations that would have seemed sad, empty, unnecessary or implausible to many of us before we experienced them, can become status quo overnight. Such novelties can also add tremendous value to our lives, even if it's not always clear whether the trade-offs result in a net win.

If the prospect of building a deep friendship with a non-human form of intelligence sounds absurd to you, consider the legion ways we already relate to non-human persons and people we've never met. Many of us routinely engage in parasocial relationships with celebrities, whose life stories we follow, invest in, and relate to from afar. We can also get deeply immersed in a great TV series, world-building video game, or burst of social media posts. And we have a profound capacity to care about fictional characters and non-human animals. Many members of my generation are, as I type this, enthusiastically raising puppies as if they are children.

Humans are storytelling animals who love getting lost in alternative worlds. But we didn't have ultra high-tech ways of generating immersive fictions in the past. So we revered what we had, engaging in flights of fancy through daydreaming, oral storytelling, the visual arts, books, magazines and theatre. Now we have Netflix, YouTube, ultra-realistic video games, virtual reality, social media, and an endless stream of novel, high-resolution pornography. Like it or not, this is not where the merger of fantasy and reality ends. Nor should we expect all aspects of that merger to be negative or dehumanising.

When we recoil at the prospect of a child forming a bond with a piece of software, we're apt to dismiss some of the valuable things that the convergence of surveillance, labour-saving and human enhancement technology has to offer. When comparing a human family with an AI companion, it's tempting to invoke idealised visions of familial bliss, in which loving parents provide their kids with everything they need to develop and be happy in life. But how often does that ideal fail to manifest, with horrible consequences? When we're sick of shelter-in-place laws during a pandemic, we're very quick to mention that many kids in first-world countries might not get enough to eat if they don't go to school, or will be stuck at home in fractious or abusive households. This is where ideals can paralyse us.

We like to think every child *should* just be able to have human parents who love and care about them, feed them nutritious food, give them the right developmental and educational opportunities and make them feel secure. But we know that many parents don't, or can't. Rates of single parenthood (overwhelmingly manifest as single motherhood) have been increasing across the developed world. One in seven families in Australia are one-parent families, while one in four children in the US lives with one parent and no other adults.[13]

Without getting too far into the weeds (and not discounting the fact that it is, of course, possible to do a wonderful job as a single parent) single-parent families are strongly correlated with worse life outcomes for children. As the Professor of Public Policy at the University of Bristol, Susan Harkness, and her colleagues note: 'single motherhood is linked to reduced income, a high risk of poverty, worse maternal mental health, poor parenting practices, and a range of other disruptions, such as home and school moves and multiple family transitions'.[14]

Attachment theory also hints at how many instantiations of the human-to-human relationships we revere, and consider gold standard, can be imperfect and damaging.[15] In their book *Attached*, the psychiatrist Amir Levine and the social psychologist Rachel Heller note that the attachment style you develop in childhood (which is shaped by the behaviour of parents and other caregivers) can have a profound impact on how you relate to romantic partners in adult life.[16] The fact that almost half of the adult population have insecure (that is, anxious, or avoidant) attachment styles, indicating some level of tumultuous and unreliable parenting, gives us further reason to think that a large minority of humans aren't doing a great job, even if they're trying their best.[17] And no parent, or family unit, is perfect.

So, what if AI could pick up some of the slack? I think AI will *eventually* do much more than this. It will be the best friend you never had. The closest thing to this deep symbiotic connection today is probably having an identical twin. But I believe the human–AI bond will ultimately become much deeper and richer than twin bonds because of the big data the AIs of the future will have access to. Unlike a human being, these non-human persons really could know you inside out.

Single parents, only children, elderly people and unpartnered adults are more prevalent today than in the past. As nuclear families cede their place as the dominant living arrangement for adults in many countries, it's becoming more of a challenge to provide people with the kind of connection that living in close physical proximity to kin in small tribes used to provide. Technology is already partially filling that void – but imperfectly, and in forms (think today's social media platforms) that are not entirely healthy. Nevertheless, we have to come to grips with the fact that there's no going back. The solutions to the challenges of the present will not be found by looking to the past – they will be found by harnessing more, and better, technologies.

One of the demographic groups that will benefit most from better AI companions in the short term is the cohort aged 65 and above. Many older and elderly people are starved of regular human connection. Aged care is expensive and often extremely poorly delivered, with little to no focus on patients' social and emotional wellbeing. People often drift into the margins of human social life as they age – kids move away, divorce is common, partners die, and some people never had anyone to begin with.

This is a problem that's getting harder to solve, even with substantial policy reform and funding. The global population is greying fast and the average number of children born to a woman in her reproductive lifetime is plummeting, which is setting us up for a future of global *depopulation*. That means there won't be enough young people to take on primary care roles, or fuel the economic growth needed to sustain aged care on a massive scale in the coming decades. The good news is that solutions exist, as you'll see in the following chapters.

Now for the bad news.

Deepfakes and post-truth

'Did you see the clip? It's amazing, I'll send you the link!' Whether it's footage of the Australian surfer Mick Fanning taking a swing at a shark as it gives chase and tries to knock him off his board, Kim Kardashian's sex tape, President John F Kennedy declaring at Rice University in 1962, 'we choose to go to the moon in this decade and do the other things, not because they are easy, but because they are hard', scenes from the Vietnam War appearing on television in America in the '60s and '70s, drone footage of rows of chained and blindfolded Uighurs in China sitting on the ground beneath police guards in 2018, recordings of George Floyd's arrest in America in 2020, or CCTV footage being used to identify potential criminal suspects, we're used to treating video evidence with a high degree of assumed veracity.

Yes, of course, there's editing and framing and the potential for tampering. Did you see the whole event? Does it look different from another angle? Is there more to the story? Is the whole thing a publicity stunt? But nobody in their right mind would attempt to argue that John F Kennedy was *not* in fact at Rice University delivering a speech on 12 September 1962, or that he never said, 'we choose to go to the moon'. And it would be the height of insanity to declare that there *never was* a man named John F Kennedy, let alone one that once served as president of the United States of America from 1961–63. That was all just fake news.

Yet I can imagine a dystopian world not too far in the future where such Orwellian 2 + 2 = 5 claims are believed, and become gospel. A world where people like Kennedy (and other such white males, who had the audacity to star in many leading historical roles in times when different cultural norms and values applied) become *un-persons*. A world where true historical archival material is drowned out by a slew of contradictory, revisionist deepfakes, created and gorged on by ape-brained humans who have either lost, or never *had*, the cognitive toolkit and skills to separate fact from fiction.

What are deepfakes? They're AI-generated content, often in video, audio, text or photographic form, which appears authentic to human users. The AI researchers at the Georgia Institute of Technology, Yisroel Mirsky and Wenke Lee, define them as 'believable media generated by a deep neural network'.[18] Deep learning is one of the subfields of AI research, which utilises neural networks, and has shown exceptional promise and extremely rapid development over the past decade. It's part of the broader subfield of machine learning and is one of the technologies behind the various iterations of DeepMind's superhuman Go-playing AIs, which belong to the AlphaGo family.

Unsurprisingly, deepfakes first captured our attention in porn. In 2017, pornographic videos were posted by the Reddit user deepfake, in which the faces of the performers were swapped with those of famous female actresses. Many deepfake videos have since been made of famous politicians, including a speech by the late US President Richard Nixon, lamenting the demise of the Apollo 11 astronauts (who, in real life, safely set foot on the moon and returned to Earth a week later). In sombre tones, Nixon eulogises, they 'went to the moon to explore in peace, [and] will stay on the moon to rest in peace'.[19] Even more astonishing is the ability of AI to generate completely realistic faces of humans who have never existed. Some of these humans have been given personas, job titles and online identities, and have been used in intelligence-gathering operations and to spread divisive messages on digital platforms.

Everybody should watch at least one deepfake video. I particularly like the pair created by the AI think tank Future Advocacy, in which the UK party leaders Boris Johnson and Jeremy Corbyn endorsed each other as candidates for prime minister in the lead-up to the 2019 general election. These videos are fun and well made, and were created to raise awareness about deepfakes. They'll probably make you chuckle. But it's also essential to know about this technology because of its broader implications. The deepfake story doesn't end with a party-trick filter that can allow you to swap your face with a friend's face in a selfie, or 'borrow' the voice of Stephen Fry or Morgan Freeman. In a few short years, deepfakes have evolved from a non-issue to a prevalent social concern, demonstrating that they have the power to destabilise societies and governments, established knowledge and history, and the ongoing pursuit of truth.

Seeing is still believing and it will be for as long as we have these ape-brains. If we see footage of Barack Obama saying, 'President Trump is a total and complete dipshit', we'll share it. Uncharacteristically, Barack Obama *does* appear to say this in a video made by the actor and filmmaker Jordan Peele – a deepfake that concludes with the fake Obama's sign-off: 'stay woke bitches'.[20] This is another video that was made to raise awareness about the power of deepfakes to distort truth and reality. It's incredibly realistic. There are hints of an uncanny valley effect when the Obama avatar's arms robotically appear to quiver, and there are slight imperfections in the syncing of his words and facial expressions. You can spot this, but you can also still readily believe you're watching the real Barack Obama.

Overnight, the possibility of content being deepfaked has given politicians a new lifeline of plausible deniability in a way that I suspect would have turned Richard Nixon green with envy. If only he could have chuckled and brushed off Watergate as fake news, as Donald Trump, Prince Andrew, and the Sao Paolo governor João Doria have all done in response to recordings or images that have surfaced in recent years, allegedly depicting them in compromising positions.[21] I'm not suggesting that denial isn't an age-old tactic. Rather, a party that is guilty of something, and chooses to deny it, may have an easier time discrediting anyone who challenges them in a climate where truth feels increasingly relative, unstable or unattainable.

If you doubt that such a climate is forming, consider the overt use of deepfakes in a 2020 political campaign by the Indian MP Manoj Tiwari. The person and the message were real: Tiwari recorded himself speaking in English. Then his team used deepfake technology to create a subsequent video of him speaking in a Hindi dialect. In the words of Tiwari's campaign manager, Neelkant Bakshi, these 'videos let us convincingly approach the target audience even if the candidate didn't speak the language of the voter'.[22]

While that application of deepfake technology might sound as benign as dubbing a film, it's a powerful symbol of the reality we're entering. We're now routinely fed a customised version of a story by search engine and social media algorithms, which are designed to appeal to, and fuel, our pre-existing biases. Reality is no longer objective or consistent; we each believe the most palatable, marketable (and increasingly, polarising) delusions that are curated just for us.

Deepfakes also have clear potential to dump gallons of fuel on the fire of hare-brained conspiracy theories, while making it harder to appear credible when exposing an *actual* conspiracy. We can cry 'fake news' whenever we don't like something, and we can share fake news whenever we do. To an ever-increasing extent, we won't be able to trust our eyes and ears any more. It will be harder to tell which images, words and audio are genuine. With more inflammatory falsehoods out there, there is also a greater incentive to produce content that's attention-grabbing enough to cut through – particularly in politics, entertainment and the media.

Now think about deepfakes in light of future defamation cases. Imagine I'm being publicly and persistently accused of being a neo-Nazi slut who eats garbage. For me to win the case I'd have to prove that this is libel or slander (not true) and lowers me in the estimation of other people. This might be very hard to falsify if video footage emerged of me wearing Nazi paraphernalia, talking about how I love the Führer, being hyper-sexual around other people, engaging in orgies, and stuffing my face with fistfuls of garbage.

Deepfakes like this could be shown in a courtroom in an attempt to prove that the bogus claims are true. Even if I could prove the footage was fake, the judge and jury would have watched it, perhaps multiple times. Remember the *continued influence effect* and the *illusory truth effect*? We often continue to believe false claims, even after they've been falsified and retracted. And our susceptibility to the *bizarreness effect* guarantees that many people will propagate a meme like this because it's sensational, lending it more credibility in the eyes of others through the *availability cascade* (keep stating a falsehood, and after it's been heard enough times, it rings true).

It's abundantly clear that our ape-brains are not designed to operate in a world where this technology is pervasive. Nor are many of the institutions and systems we created in simpler times, in domains like law, journalism, education and governance. But the technology's not going anywhere. It's improving, fast. This leads us back to the conclusion I've been affirming throughout this book: *we're the ones who need to change.*

I'm not suggesting that misinformation is a new phenomenon, or that our ancestors didn't believe (or weren't fed) a whole lot of codswallop. But human civilisations have come a long way since the scientific revolution. Robust knowledge accumulates faster, and mass literacy enables more critical, independent thought and analysis, and greater freedom of expression. My worry is that humanity may permanently forfeit some of these hard-won gains if we languish too long on the threshold of a post-truth era.

As a general principle I think it's a bad idea to latch on to a present phenomenon and worry that the future will be dominated by that lone risk or reward amplifying itself. It's never that simple. But I've got a tad locked on to the issue of misinformation, because of how much of what we need to safeguard the future – science, reason, and a respect for empiricism, truth and free inquiry – stand to be eroded and trampled on. And not just from fake news and deepfakes, but also by one of the related trends I was really hopeful about a few years ago.

Human capital, connectivity and the last three billion

I was totally enamoured with the tech-saturation narrative in the early 2010s. I thought bringing the last three billion humans online would have a kind of magical ability to lift people out of poverty, connect them, give them more stable digital currency (anyone can steal your goat, but it's much harder to hack your bank account) and expose them to a bigger world of ideas and aspirations. I still think there's a lot of truth to that, and would never in good conscience argue that such technology should be withheld. I believe in freedom of speech and freedom of information and I still have a soft spot for the mantra from the early days of cyberspace: 'information wants to be free'.

But we've been learning more about this global experiment as it's played out. The internet initially looked like it might break down barriers and divisions between people of different ages, countries, religions, social classes and cultures. But the internet of the 1990s was very different to the internet of today. For starters, very few people used it. Guess what kind of people were disproportionately getting their hands on dial-up modems while others rolled their eyes? Very smart, educated and tech-savvy ones. Nerds, you might say.

The early net wasn't a digital utopia: it was full of porn, hackers gleefully circulated computer viruses, and bullying and sexual harassment definitely took place. But guess what else happened? Lots of conversation. People wrote stuff down, shared information on websites, emailed, posted on bulletin boards and regularly exchanged ideas. The kinds of people who enjoyed this were (surprise, surprise) disproportionately educated, literate and intellectually curious.

There was also *much* less support for image and video content. We're used to thinking of the web as a visual feast, but that's a very recent thing. I can remember a tipping point in the last decade where blogging suddenly went from 'maybe throw a few pictures in there' to: 'IT'S ALL ABOUT THE HEADLINE AND THE COVER IMAGE'. Add high-quality smartphone cameras and social media into the mix and digital self-presentation is no longer about how you write or think, it's about how you look and which slogans you chant, tweet and endorse.

A more democratised internet means user demographics no longer skew towards smart outliers, they follow the bell curve instead. This results in less average intelligence at the digital town hall. Does a rising tide lift all boats? Or does a strong regressive current make ships sink? I wonder about this when I think about the last three billion humans joining the global digital brain.

I think universal internet access will be a net gain for humanity, and coming online will be a profoundly beneficial step for most individuals. But I still fear some of the potential costs of the short-term disruptive phase, during which billions more humans suddenly appear online. Many among the last three billion will surely discover online courses, and access some of the world's best information, and plenty *will* be lifted out of poverty. But I'm concerned that an alarming proportion will be driven towards the same kinds of superficiality and tribalism that the rest of us have embraced, adding billions more (suddenly audible) voices to the anti-Enlightenment and anti-science chants we're hearing more of these days.

Look at what's happened to the nearly five billion of us who are online. We reached into a digital utopia where almost anything could be possible and we pulled out the African savannah. We siloised. We built echo chambers and gravitated to the ancient norm of pitting tribe against tribe. High-bandwidth connectivity brought all the visual, superficial cues of human difference back to the forefront of cyberspace. At the same time, many of us now glorify, or aspire to emulate, the crude amplification of ancient cues of fertility and fitness under the flawed principle of bigger is better, taken to ever-greater extremes – Kim Kardashian and the legion self-made Instagram models who've emerged in the past fraction of a second of human history are a case in point.

As radically different cultures and communities suddenly converge and collide online, what was once quietly inoffensive and happening halfway around the world turned into something that must be named, shamed and silenced. Instead of live and let live, we want to conquer, expunge, scapegoat and pillory. When I see statues of yesterday's elites and flawed historical figures falling to the gleeful chants of a mob, I cannot help but think of a passage from Milan Kundera's *The Unbearable Lightness of Being*:

> She would have liked to tell them that behind
> Communism, Fascism, behind all occupations and
> invasions lurks a more basic, pervasive evil and that
> the image of that evil was a parade of people marching
> by with raised fists and shouting identical syllables in
> unison.[23]

There are more bad ideas than good ones in the world and by connecting with billions of other humans, we've gained daily exposure to far more bad ideas than most of us would ever encounter just going about our lives in the analogue world. That doesn't necessarily mean that a greater proportion of extreme nutbaggery exists than it used to. But we *see* it, which triggers tribal impulses that make us want to react and fight it. When we do so, we are often fanning the embers of a memetic flame that would probably have died out if we just left it alone.

No primrose path

There is a degree to which AI is a hail Mary for humanity. I would love to be proven wrong about this. I really hope we can design its seed programming carefully enough to maximise future benefits and avert catastrophes as its capabilities and influence grow. But there are so many unknowns that lie ahead in our co-evolution with artificial intelligence. It's pretty hard to imagine things working out perfectly at every stage – especially as we develop new products and systems that transform our world faster than we can predict their long-term effects. Kevin Kelly, whom we met at the beginning of the chapter, put it well when he wrote, 'We are morphing so fast that our ability to invent new things outpaces the rate we can civilize them'.[24]

The AI safety researcher Eliezer Yudkowsky also reminds us of what can go wrong when it comes to developing strong AI that matches and then exceeds human intelligence:

> If we could just try over and over until we got it right, and we had 200 years to get it right, I'm very confident we could get it within two centuries. But the real world isn't like that; in the real world, if you get it wrong on the first try, you die. And if you take too long, someone else just goes ahead and builds an AI. That's what makes the problem difficult; the good guys have to get it right on the first try, and there are various other people tackling the intrinsically easier problem of building an AI regardless of whether it's safe.[25]

I think Yudkowsky's right. The trouble is, at this point in human cultural evolution, AI is also an essential tool that we will need to utilise in order to solve the growing number of problems that are already on our plate – including helping us to identify and filter out the very misinformation and brain junk it's playing a role in creating. We hear so much about the human cost of rapid technological growth in books like *The Age of Surveillance Capitalism* by Shoshana Zuboff, and *Zucked* by Roger McNamee. We hear much less about the human cost of not upgrading, and failing to develop technologies like human-level artificial intelligence and superintelligence.

As you'll see in the following chapters, we can't maintain an early-21st-century status quo without enlisting more help from our advanced technological friends. Smarter technologies offer the only way out of the demographic, economic and environmental crises we're barrelling towards. We need automation to fuel the economic growth necessary to keep innovating and driving the price of clean energy solutions down. We need automation to generate enough wealth to sustain a rapidly ageing global population. Before long, there won't be enough young people to meet the world's productivity needs. And without better robots and AI, living standards will decline and social turmoil will rise.

If we're going to be locked in to a world of automation-driven abundance, we'll also need advanced technologies to create new forms of purpose and pleasure – and to extend our healthspans and lifespans, because we're not replacing each other fast enough through biological procreation. That might sound dire, but here's the 'galaxy-brain' version of how to think about humanity's impending demographic crunch.

The post-industrial population boom paved the way for our species to reach a new level of technological maturity, facilitating the invention of computers, the internet and artificial intelligence. Digital intelligence evolves much faster than biological intelligence and could out-evolve us in the blink of an eye. Digital minds can make copies of themselves and proliferate on a scale that defies the limits of human population growth. *Human* populations will contract, but digital forms of intelligent life could proliferate on our planet on a scale never seen before – leading us to that event horizon of a posthuman (and hopefully superhuman) future.

While there are many ways the future could play out, I can readily imagine human-level AI and superintelligence doing away with a lot of things we value now, and causing some human suffering, while bringing massive net benefits to humans and our successors in the long run. This is one of the better possible futures that humanity should aspire to. But that's a very hard prospect to be relaxed about if you're a human who might lose some things you value today, or tomorrow – and I think that basically describes us all.

AI is going to be a major catalyst for many of the weird and wonderful changes you can expect to see in your lifetime. As its capabilities and influence grow, it will ultimately pose a greater existential threat to humanity than nuclear weapons.[26] That might sound frightening, but there's no closing the lid on the Pandora's Box that contains AI. Which is probably a good thing, because we're *guaranteed* to lose it all – humanity, our habitable planet, and the evolutionary potential of intelligent life – if we don't invent something smarter than ourselves.

In the meantime, I'd urge you to keep Kevin Kelly's prophecy in mind when thinking about what to study, what kind of business to start or invest in, or what kind of life arc you imagine unfolding. It's time to relinquish our 20th-century expectations, and start betting on a future that's going to be much stranger than you think.

8

LIVE FOREVER
OR DIE TRYING

Give me health and a day, and I will make the pomp of emperors ridiculous.

Ralph Waldo Emerson, *Nature*

Any death prior to the heat death of the universe is premature if your life is good.

Nick Bostrom, 'Letter From Utopia'

Most people who are older than I am tend to scoff when they hear that this kid of 30 is hankering to look, feel and if possible *be* younger. 'Oh, you're still a *baay*-bee', they say with a knowing grin. Maybe according to your eyes, but not according to science. After the age of about 20, your risk of succumbing to major age-related diseases increases exponentially, so I'm already a member of the slowly (then suddenly *very* rapidly) falling apart club.[1] Nice to see you here, and my sincere commiserations.

If historical precedent and current life expectancies are anything to go by, I can confidently bet that in eighty-five years' time, the human who is currently tapping away on the black and white keys of a Macbook Air, thinking excitedly about the future, and talking to you all in this clunky form of telepathy, will no longer be among the living. I will be no more, have ceased to be, expired, gone to meet my maker, shuffled off the mortal coil and become a decidedly *ex-human*!

In current circumstances I've got a *one in one hundred million* chance of making it to the age of 115, and given the current quality of life of someone that age, it's a privilege I'd choose to forgo.[2] But what a terrible choice: to either live longer and watch yourself decay, or take your curtain call and exeunt. I'm not pleased with this suite of options and I doubt you are either if you're being honest.

Thankfully, I'm not the only oddball dwelling on this particular gripe. A growing group of scientists are now vocally campaigning to declare ageing a disease. Along with technologists, philanthropists and government bodies, they have the skills, potential, cash and power to try and get us out of this terminal mess.

Strategies for Engineered Negligible Senescence

For most of human history we've thought of ageing as inevitable. But today's scientists are starting to approach it more like an engineering problem – something to be mapped, hacked and solved. By exploring the mechanisms behind ageing, which are the main factors that lead to us getting sick and falling apart, they hope to usher in an era of preventative and personalised medicine, targeting many diseases, ailments and chronic health conditions at their source.

One of the most recognisable figures in this mission is Aubrey de Grey. A software and artificial intelligence engineer by training, de Grey became interested in the convergence of biology and technology in the late 1990s and began independently researching the biology of ageing, with mentoring from his biologist wife (now ex-wife) Adelaide Carpenter. This work culminated in his book *The Mitochondrial Free Radical Theory of Ageing*, for which he was awarded a PhD in biology from the University of Cambridge in 2000. De Grey went on to become the Chief Science Officer of the SENS Research Foundation, a non-profit dedicated to funding research and promoting access to rejuvenation biotechnologies that can bring human ageing under control.

At present, it's believed that there are about nine major processes that contribute to ageing, including: DNA damage and cancer-causing mutations, epigenetic changes, mitochondrial dysfunction, the shortening of telomeres, senescent cells that accumulate in our bodies, the dysregulation of protein function and nutrient sensing, the depletion of stem cell reserves, and impaired cellular communication.[3] De Grey believes that by tackling them concurrently, we will be able to continuously repair and rejuvenate damaged human cells and body parts, much like a car that has its parts maintained and replaced decade after decade.

Let's take the example of senescent cells. You'll often find these around tissue that's ageing and they're associated with inflammation and dysfunction. Senescent cells are old cells that have stopped dividing and have attained a zombie-like state. But instead of being cleared away, by a process called apoptosis, they hang around and secrete various nasties that damage nearby cells, contributing to the development of many age-related conditions. The good news is that we could reverse the damage and rejuvenate the body by deleting these old cells. Work at the Mayo Clinic by the biochemists and gerontologists Jan van Duersen and James Kirkland led to the pioneering of what are now called *senolytics* – drugs that can eliminate senescent zombie cells without damaging neighbouring cells.[4]

The exciting part is that senolytics are just one of many tools – including better drugs of every stripe, gene therapy, stem cell therapy and big data–enabled preventative healthcare – that could help us combat the ageing process and remain vital and healthy for longer. The fable of the long-awaited magic pill is exactly that – a fable. Ageing is a multi-factor process and it manifests as damage, dysregulation and decline. Tackling the causes of ageing in concert amounts to what de Grey refers to as Strategies for Engineered Negligible Senescence (SENS). In other words, hacking your body so that the amount of age-related damage that accumulates over time is basically nil.[5]

Eventually, de Grey thinks humans will achieve something called Longevity Escape Velocity (LEV), which is the idea that human life expectancy will increase by more than one year every year. Although LEV would not eliminate death by accident or mishap, the idea is that it might give those alive a real chance of attaining radical longevity, as further progress occurs in the fields of medicine and biotechnology. The longer you live, the better your chances of accessing a suite of more powerful preventative and regenerative therapies, which are expected to emerge at an accelerating rate.

De Grey's expectation of accelerating progress in bio-technology, which underpins his belief that human lifespans could be *radically* extended for thousands of years, has been controversial among biologists, to say the least.[6] But the broader scientific aspiration to understand and treat ageing as a multifactor disease has swiftly transformed from the province of kooks on the fringe to something that increasingly attracts funding, is pursued by biotech companies, and is discussed in leading medical journals like the *Lancet* and the *Journal of the American Medical Association (JAMA)*.[7] If nothing else, de Grey is fighting a battle to normalise the idea that death within a century (give or take) is not necessarily inevitable, and should not simply be accepted with a sigh as 'the way things go'.

Why do humans age and fall apart within a century anyway?

A key reason is because the force of natural selection declines with age — so it's not as good at weeding out late-acting harmful mutations as those that affect us early in life.[8] It's rare that we experience debilitating conditions before we hit reproductive maturity because genes that kill us before we pass them on won't survive. But genes that start causing problems later in life will have a much easier time hitching a ride down the hereditary road. As we age, we feel their deleterious effects with increasing force.

The good news is that we only have to look at other species to know that the precise trade-offs incarnate in human bodily designs are not the only biological options. The naked mole rat appears to be a remarkable outlier among rodents and has been the subject of some very interesting research at Google's biotech subsidiary Calico. The species lives five times longer than similarly sized rodents, has over a thirty-year lifespan in captivity and may even defy Gompertz's law, which describes a phenomenon where the risk of death increases exponentially with age – as it does for humans and the rest of our mammalian kin. Breeding females do not experience menopause, and retain their fertility past the age of 30; incidences of cancer are extremely rare; and no significant changes in body composition, metabolism, bone quality or cardiac function are apparent over several decades. Cellular function and protein expression also appear to remain robust with age.[9]

Some animals also rejuvenate far more effectively than humans, like the axolotl, and the colloquially named 'immortal jellyfish'. At any age, axolotls can regenerate their tails, limbs, eyes, lungs, spinal cords and even parts of their brains and hearts several times. In 2018, the enormous axolotl genome (it's ten times larger than a human's) was sequenced for the first time.[10] Researchers are now busy looking for clues that explain the mechanisms behind this amazing ability, which might help advance the progress of regenerative medicine.

The immortal jellyfish also has the incredible ability, under stressful conditions, to transform from a mature adult back into a polyp, before maturing again, and so on. Essentially, it can age both forwards and backwards. It's also worth mentioning the curious phenomenon of Peto's paradox. This is named after the epidemiologist Richard Peto, who wondered why larger animals – which have more cells and therefore a higher theoretical chance of experiencing more copying errors, and cancer-causing mutations –

don't succumb to cancer at much higher rates than smaller animals. Humans have many more cells than mice – about 1000 times more – and we live thirty times longer. Yet our risk of contracting cancer doesn't differ substantially from that of a mouse.[11]

It's also curious that some *very* large animals, like elephants and whales, experience remarkably low incidences of cancer. There are many possible explanations for this. Some animals, like the African savannah elephant, appear to have evolved superior tumour-suppression mechanisms. The elephant's genome contains twenty copies of the tumour-suppression gene TP53, while the human genome contains only one. Discussing Peto's paradox, the biologists Marc Tollis, Amy M Boddy and Carlo C Maley highlight the fact that 'gigantic animals do not get more cancer than humans' and believe it suggests 'that super-human cancer suppression has evolved numerous times across the tree of life'.[12] So, why not focus on figuring out how, and try to replicate some of those advantages in humans?

The more evidence you find in nature of superlongevity, and superhuman rejuvenation capabilities, the more plausible it is that we can devise solutions to at least some causes of human ageing and frailty. Of course, humans aren't axolotls and there are many trade-offs in our bodily design. It's rare that you can make tweaks without unintended consequences. But that's where more computing power and better AI stands to play a major role – in helping us create ultra-precise, complex and dynamic maps of every part of the human body. If we modulated many aspects of human design in concert, we could cultivate a more robust system than the one we inherited from our blind evolutionary architect.

Proofs of concept

There are also many precedents for hacking the ageing process in the lab in other species. One way to do this is to modify lifestyle or environment. For many species, cold environments and reduced body temperature appear to promote longevity.[13] Caloric restriction (consuming substantially fewer calories, without triggering malnutrition or starvation) has also been shown in thousands of studies to extend the lifespans of many organisms, from yeast to worms, mice, rats, dogs and primates.[14] In humans, caloric restriction lowers risk factors for major diseases, improves longevity biomarkers, and is correlated with longer lifespans.[15]

Many drugs and supplements have also been shown to improve lifespan and healthspan in model species used to study ageing. Promising substances include rapamycin, metformin, resveratrol, nicatinamide mononucleotide (NMN), acarbose, and 17-α-estradiol. Rapamycin is an FDA-approved immunomodulator, which has been shown to extend the life expectancy of mice by up to 60 per cent, while improving the healthspans of middle-aged mice and reducing their rates of cancer and cognitive decline.[16] Rapamycin isn't approved to treat ageing in humans (no drugs are, because ageing isn't considered a medical condition) but it was shown to improve the immune response of elderly volunteers to an influenza vaccine by 20 per cent, raising the possibility that it could be used to target immune system decline.[17]

Researchers are also studying the role of the microbiome in health and longevity. Middle-aged African turquoise killi-fish that received a gut bacteria transplant from younger donor fish were shown to live longer and experienced a delayed rate of behavioural decline.[18] Pro-longevity results have also been achieved by injecting the plasma of younger rats into that of older rats. Researchers used epigenetic clocks (aggregate biomarkers of age) to measure the effects on the tissue and organs of the older rats. Compared to a control group, old rats treated with plasma from younger rats experienced an average of 54 per cent reduction in their molecular age across tissue types, more than halving their epigenetic age. The treated rats also exhibited improved memory and cognitive performance and a reduction in their levels of inflammation and cellular senescence.[19]

Many longevity pathways have been conserved in species across deep evolutionary time.[20] That's why animal studies are useful, even though they don't always predict results in humans. They can help us identify which genes and pathways to study and test. With large-scale genomic sequencing studies, and AIs mining those datasets, we'll be able to map universal features across species and study them with far greater depth, precision and rapidity.

There's also the ever-tantalising prospect of gene therapy. In 1993, the biochemist Cynthia Kenyon identified genetic mutations that enabled the *C. elegans* roundworm to live 50 per cent longer.[21] Since then, thousands of lifespan-modulating genes have been identified in model species and tweaked to extend their healthspans and lifespans. Many of the same longevity genes have also been identified in humans.

Not coincidentally, researchers who study ageing are very interested in exceptionally long-lived humans – centenarians (or if they make it past the age of 110, supercentenarians). In his book *Age Later*, the gerontologist Nir Barzilai tells the story of the Kahn siblings – Irving, Helen, Peter and

Leonore – all of whom lived in a state of remarkable independence past the age of 100 (Irving was still working at the family investment firm in Manhattan at the age of 108). The siblings didn't have unusual lifestyles, or adopt any kind of special diet. Helen, who died at the age of 110, smoked for more than ninety years. When asked if any of her doctors had ever advised her to quit, she replied, 'Sure, but all four of those doctors died'.[22]

What's a centenarian who smokes all her life, without it doing her any harm, got that the rest of us don't? Great genes, apparently. Researchers are turning the powers of genomic sequencing and AI on to families of super-agers, in the hope of unlocking the mysteries of pro-longevity pathways – which nature has already shown to exist in some humans.[23]

Fighting ageing as a disease

David Sinclair is one of many scientists who believes that human ageing will be conquered this century, and that the condition should be treated as a disease. Sinclair began his studies in molecular genetics in Sydney in the late 1980s. Today he's the co-director of the Paul F Glenn Center for the Biology of Aging at Harvard Medical School. On many occasions, he's declared that he's confident the first human who will live to the age of 150 or beyond is alive today.[24]

In his 2019 book, *Lifespan: Why we age – and why we don't have to*, Sinclair argues that current approaches to treating age-related diseases are limited and costly, as they focus on treating symptoms and diseases one at a time. He points out that:

The United States spends hundreds of billions of dollars each year fighting cardiovascular disease. But if we could stop all cardiovascular disease—every single case, all at once—we wouldn't add many years to the average lifespan; the gain would be just 1.5 years. The same is true for cancer; stopping all forms of that scourge would give us just 2.1 more years of life on average, because all other causes of death still increase exponentially.[25]

The British geneticist and biogerontologist David Gems, who is the Deputy Director of University College London's Institute of Healthy Ageing, helps explain this phenomenon further, remarking:

> The battle with aging is akin to that between Heracles, the hero of Greek mythology, and the multiheaded Hydra. Each time Heracles hacked off a head, two more would sprout in its place. Likewise, the old man successfully treated for prostate cancer may not long afterward stagger back into the physician's office with macular degeneration and dementia. Such piecemeal approaches to treating age-related illness have undoubtedly improved late-life health to an extent and they have increased life expectancy. This, again, is something to celebrate. Yet in the long run a more powerful way to protect against age-related disease would be to intervene in the aging process itself. This would provide protection against the full spectrum of age-related illnesses. Returning to our classical illustration, to really defeat the diseases of late life we

need to strike at the heart of the Hydra of senescence: the aging process itself.[26]

This once-fringe perspective is gaining ground in the biomedical community. In 2018, the World Health Organization added an 'Old age' extension code to their disease classification manual (the ICD-11), which the editors of the *Lancet* described as representing 'undeniable progress towards overcoming the regulatory obstacles that have thus far hampered the development of therapeutic interventions and preventative strategies targeting ageing and age-related diseases'.[27]

While this is a step in the right direction, the next step is for ageing to officially be declared a disease. The classification matters for a few reasons. When it comes to ageing and medicine, there are currently no standardised metrics about when ageing should occur, what biomarkers should be apparent, and how fast it should progress. Hence the truism 'everybody's different'. Sure, but *how* are we different and *why*? Why do some people never get sick a day in their lives and live into their 100s, while others fall apart in middle age? How do we unravel the complex interplay of genes, bodily processes and lifestyle choices, so we can identify optimal combinations and more evenly distribute the benefits of health and longevity?

We need to study the many systems and processes that contribute to ageing. As the CEO of the AI-focused drug-discovery company Insilico Medicine, Alex Zhavoronkov points out, it's much easier to study something when you have a disease classification. The historical precedents of obesity and autism, currently both classified as diseases, indicate that a disease classification results in 'the development of more accurate diagnostic methods, and increased involvement of the pharmaceutical industry and policy makers'. It 'also provides the basis for clinical trials'.[28]

Drug companies won't invest time and money targeting the causes of ageing if they're not allowed to sell anti-ageing drugs. Currently, drugs like metformin, which have shown promise in increasing human life expectancy, can't be prescribed for the condition of 'ageing' because ageing is considered medically normal. Metformin is a cheap drug that has been widely prescribed to diabetics for decades and is also used to treat women with the common condition of polycystic ovarian syndrome (PCOS). Remarkably, diabetics who take the drug have lower incidences of cancer and live longer than matched controls without diabetes who do not.[29]

That information basically fell into our laps, because we'd already invented this drug for other purposes and had a large cohort of people taking it, whose life outcomes we could monitor under the rubric of diabetes research. It's possible metformin only has longevity benefits for those who are already metabolically unhealthy – but there's only one way to find out. The good news is that the American Federation for Aging Research (AFAR) has recently launched a large-scale human clinical trial – known as the Targeting Aging with Metformin (TAME) trial – to study the effects of metformin on the progression of age-related chronic diseases.[30]

If the results show promise, the FDA will likely approve future trials designed to study *ageing itself* as a target for drug discovery and development – as opposed to single, age-related conditions like heart disease. That's a great step forward. But think of all the anti-ageing therapies we might have invented (or existing drugs we could have repurposed) if we'd actually bothered to look for them in decades past.

The problem of hype

My Italian grandfather is 94, and he never misses an opportunity to tell me that radical medical advances are codswallop. They promised they were going to cure cancer fifty years ago and they still haven't done it. That's certainly the conclusion you'd draw if you had a questionnaire in front of you that read: *has humanity cured cancer?* There are two boxes – one labelled Yes, and the other labelled No. You can only tick one. Of course we're going to have to tick the No box. But multiple-choice questionaries are hardly renowned for telling nuanced stories.

In many respects, the gains humanity has made in the biomedical space over the last century are *under*hyped. While more people are diagnosed with cancer today than in the past, that's because there are more people in the world, and more of us are living longer. The headline we don't often see is that survival rates for most common cancers have increased dramatically over the past few decades, thanks to lifestyle modifications, screening programs, early intervention and better treatments (including the promising emerging avenue of immunotherapy).[31] HIV is no longer a death sentence.[32] Global child mortality has declined fivefold since 1950.[33] And polio has been eradicated in every country in the world, barring parts of Afghanistan and Pakistan.

We have a tendency to assimilate these gains into our new sense of normal, and dismiss them as no big deal. But although we tend to underhype certain markers of human progress, that doesn't mean *over*hype isn't a problem too. I think it's fair to say that some of the most visible proponents of anti-ageing science focus excessively on the promise of their field, while remaining cagey about some of the challenges and possible impediments to progress. Getting people excited about promising biomedical research is great. People *should* be excited about this (much more than they typically are) as there is real potential for incredible advances in this arena. But the details are complex, the timeframes unclear, and the possible roadblocks legion.

I suspect that longevity science leaders like Aubrey de Grey and David Sinclair each believe passionately in the goal of extending human lifespans and healthspans and reducing human suffering. I also think their arguments in favour of *pursuing* healthspan and life extension make a tremendous amount of sense. These men are both highly intelligent, and are well versed in the science of ageing. But they're still human, and it's hard to set yourself up as a 'guru who's not like other gurus' without ultimately becoming a lot like other gurus.

I interviewed Aubrey de Grey by email when writing this chapter. He was very generous with his time and typed out some lengthy responses to my questions. But they were the kind of responses I'd read in interviews with him dozens of times before. I think Longevity Escape Velocity is a logical prospect, and de Grey's popularisation endeavours have helped bring the urgency and viability of life-extension research to the fore. But after some pressing and probing, I found I couldn't get him to engage with a discussion about points of failure. What assumptions might we have encoded into life-extensionist hypotheses that are incorrect? What hard problems in biology might stonewall us unexpectedly, at least for a time?

This is where I think the engineering mindset (the determination to solve a problem, and to build new tools to fix things) might benefit from coming back into contact with a scientific mindset – which looks beyond the goal, and considers our favourite hypothesis (that we'll be able to achieve radical life extension in humans) alongside other competing hypotheses, including the prospect that we might struggle to achieve it.

De Grey fascinates me, because at a glance it looks like his main job is to hype an idea into becoming a self-fulfilling prophecy. To be clear, he's an exceptionally smart man, who's done genuinely interesting research in biology and does things like solving 'unsolvable' problems in the field of mathematics in his spare time.[34] But he's also successfully created an accessible narrative about life extension (in the form of SENS and LEV) that's helping to spread the word and attract funding for life-extension research.

The more successfully you generate funding, the more likely it is you're helping speed up breakthroughs, while turning the tide of cultural perception from one of hostility to one of enthusiasm. I often wonder to what extent de Grey is consciously using hype as a promotional tactic. In many ways it's a smart strategy. But I don't think it's helpful to avoid nuanced discussions about roadblocks, even if the ultimate justification is that it would require getting into territory that's too technical. It's important to get the balance right between promoting an idea or cause, and effectively communicating key facts, rigorous research and nuanced details.

Where Sinclair is concerned, that balance seems to have skewed more in favour of public hype in recent years – perhaps not coincidentally as the social media era has ramped up. Social media is such an irresistible and effective tool for sharing exciting preliminary results of an animal study, or discussing the big-picture potential of reseach in a new preprint. But Sinclair can't possibly be unaware that he has nearly 200 000 Twitter followers, many of whom are desperate to hear 'the science says cures for all my ailments are coming', and who seem to particularly want to know which experimental molecules he personally ingests to promote his own longevity, and in what doses.

In many Facebook groups and online forums, he's treated like a guru and a god. 'What does Sinclair take? How much resveratrol, quercetin, NMN?' In many cases, these followers aren't asking so that they can go and do a bunch of careful thinking and independent research, and determine if that dose might be appropriate. I often wonder how many of these avid self-optimisers have carefully considered the potential risks and benefits of ingesting substances that have not been taken by humans long term, or extensively studied in clinical trials. Many appear to simply want to copy his formula – and for the most part, they do.

I'm genuinely torn about this. I think sharing preliminary research findings can be very useful, and people should be free to do independent research and decide what they put in their own bodies. But freedom to decide, and the capacity to decide *well*, are two very different things. Those who know they have profound influence over others should, in my view, be very careful in how they choose to exert that influence.

I'm not a biologist. I don't research ageing for a living. But I do follow this space keenly and have for a good decade. I sympathise with the desire to promote a field you believe is truly important, particularly when the study of ageing has spent so long being considered a 'backwater' of science.[35] But I also think overhype can damage the credibility of an endeavour, particularly if research avenues, or potential therapeutics, fail to yield the hoped-for results – as many undoubtedly will.

For my part, I haven't written this chapter in order to promise you anything. I'm in no position to do so. My aim is to highlight a vastly underappreciated point about the pursuit of better health, and greater longevity – both are *possible*, both are desirable, and they're goals we should be actively pursuing. Here's a helpful way of thinking about why.

Imagine you're 30 years old and the year is 1922. What probability would you assign to being alive and healthy (as in, not suffering from chronic disease and disability and living as a shadow of your former self) in a hundred years? I'd put the probability at 0 per cent. I know we have the benefit of hindsight here, but that's also what most people in 1922 would have said. Now run the same thought experiment for 2022. It's *very* hard to say 0 per cent with a straight face. There are too many ambitious projects in the works, and too many advanced technologies recombining to accelerate gains in the fields of biotech and biomedicine.

Many of the world's biggest tech companies, and richest humans, are actively investing in conquering ageing, disease and involuntary death. I'd put the chance of a 30-year-old today being alive and healthy (or potentially digitally instantiated, which I'm counting as 'alive') in the year 2122, at around 10 per cent. You might quibble with that number, but what matters is that it's very unlikely to be zero. We instinctively discount what a huge deal that is, because our dominant expectation is that we'll be dead in a hundred years' time. Yet even if we are, we have a better shot than any of our ancestors at dying later, ageing more humanely, and dying well.

The disruptive phase of change

Of course, we all want to know what a regular trip to the doctor will look like in ten, twenty, or fifty years' time. I asked Aubrey de Grey about this and he replied: 'Not very different from today, except in your motivation: it won't take becoming sick to make you go there, because you'll have understood that preventative maintenance is the way to go. The treatments themselves will be very much like today's from the patient's perspective: injections, mainly'.

That sounds anticlimactic, but most profound interventions in human history do. The American agronomist Norman Borlaug didn't stage a dramatic hunger strike or nail himself to a cross to save lives. Nor did he shoot any bad guys, or stage a rock concert to solve poverty. He spent a lot of time thinking about the microbiology of wheat. After decades of work, he successfully bred new, high-yield, disease-resistant strains of wheat that helped avert the much-feared global food crisis that many were adamant would materialise in the mid-20th century. Bourlag is credited with saving somewhere between 10 million and a billion lives.[36]

'Injections, mainly' sounds banal. But de Grey is talking about injecting senolytics that knock out old cells, stem cell and plasma injections that rejuvenate us from within, and gene therapy that targets the cause of illnesses at the source. We're back to the same problem of narrativising the future that we talked about in the introduction. If I told you the most hyped-up version of this story, I'd rattle off a list of companies offering cool new products and therapies that promise to make us live longer, healthier lives. There really are some amazing things in the works – from a 'universal' blood test that can detect the presence of more than fifty types of cancer, to the next generation of vaccines that could target malaria, HIV, cancer, or multiple respiratory viruses simultaneously, and 'organs on chips', which allow researchers to recapitulate, and study, the living microarchitecture of body parts including lungs, livers, kidneys, skin, bone marrow and the blood–brain barrier, noting how they respond to drugs without having to rely on animal studies, or be impeded by the difficulty of viewing these body parts *in vivo* in humans.[37]

AgeX Therapeutics is a company focused on stem cell therapy and hopes to bring products for tissue and organ rejuvenation to market. If you have a damaged heart, or spinal cord, or eyesight fading from macular degeneration, the goal is to provide you with a localised injection of young stem cells, which can repair and regenerate those body parts. Meanwhile, LyGenesis, Inc is hoping to bring an end to the donor organ shortage by using lymph nodes as bioreactors to grow new organs. They're starting with livers, and hoping to cure end-stage liver disease, which currently requires a $700 000 transplant procedure to treat in the US, if you're lucky enough to find an organ donor. The LyGenesis team hopes that their ectopically grown organs could eventually be transplanted with a minimally invasive endoscopic surgery.

The trouble is, most biotech startups fail and most of the products scientists work on don't live up to the hype. But that doesn't matter if you're developing millions of products and investigating millions of different therapeutic strategies – and using AI to accelerate the process to warp speed. Don't forget, Moderna's Covid-19 vaccine was designed in a weekend, only days after Chinese scientists first shared the gene sequence of the SARS-CoV-2 virus.[38] If only 10 per cent of humanity's biotech developments have benefits, you'll still end up with hundreds of thousands of great new things – and many products that initially fail will still be based on ideas that have promise.

Whether or not Elon Musk's company, Neuralink, ultimately becomes the behemoth that pulls ahead in the implantable medical technology space, think about the core product: a Bluetooth-enabled chip that can continuously record brain activity. And what about scaling less invasive existing tech, like EEG wearables? AI mining millions, or billions, of continuous streams of data coming from our brains is going to unearth interesting patterns that massively accelerate the pace of research, and the number of breakthroughs in neuroscience. The time is ripe, because the technology is finally catching up to humanity's grand ambitions to map, hack and enhance our abilities to superhuman levels.

One of the biggest killers of young people is suicide. If we finally understood the forms of brain activity that contribute to depression, we could target those areas of the brain with precision drugs that actually worked – as opposed to current classes of antidepressant that sort of work for some people some of the time, and we don't really know why. In other people, these drugs may actually increase suicidal tendencies. It's a dice roll, which epitomises our current status quo of *imprecision* medicine.

With sophisticated AI companions, our speech could be continuously monitored for slurring, changes in timbre, and other cues of inebriation, illness and mental health problems. On longer time scales, machine–brain interfaces, and even nanobots that are injected into our tissue or bloodstream, could not just monitor, but repair us in real time.[39] Meanwhile, stem cell therapy has the potential to help us regrow genetically compatible replacement tissue, blood vessels and even organs. We're already at the dawning age of personalised medicine. The next step on the horizon is regenerative medicine, where we repair damage while preventing as much of it as we can from occurring in the first place.

The French philosopher, scientist and mathematician René Descartes thought medicine in the 17th century was mostly bunk – and he was right. Of course, it's easy to look back and chuckle at their use of bloodletting, and silly rhetoric about balancing the bodily humours, or the soul residing somewhere in or around the pineal gland. But we don't often think about how ignorant we will seem to future beings on the medical front – even to our future selves.

Soon, we will look back on this period of normative healthcare standards and knowledge and be stunned by how little we knew, and how crude our tools were. We will be flabbergasted by how pervasive and normalised widespread human suffering was. In a sense, we are living in the dark ages, but we won't be aware of this until we start to see the light.

Your government wants you
to stop falling apart too

It's rare that governments chase moonshots and as far as I know, none are overtly making radical life extension a core priority. But in their own ways, many are pursuing it because it's the incidental result of tackling a problem of crucial significance to more than half the governments around the world, including the US, China, Japan, Australia, Korea, New Zealand, and all of Western Europe. That problem is old people. We have a lot of them and they're very expensive.

Why so expensive? Because they're literally falling apart and need lots of healthcare and maintenance. Most of them don't work any more and they're drawing from, rather than paying into, the tax system. Governments don't tend to worry much about the humanitarian crisis of ageing and its impacts on the lives and psychology of the elderly and their families. But they do worry about the bottom line.

We're heading for a future where there won't be enough young people to support the enormous cadre of ageing folk who want, and deserve, years or decades worth of medicine, aged-care facilities, pensions, and a reasonably dignified stint in God's waiting room. According to the World Health Organization we're going to see an almost doubling of the proportion of the world's population aged over 60 years between 2015 and 2050.[40] We crossed a telling threshold in 2018, when the world found itself home to more adults over the age of 64 than children under the age of 5 for the first time ever.[41]

Kids are rapidly disappearing from our world, but we haven't noticed because we invest much more in the ones that are here and we're getting more used to childless adults and smaller families. As long as some people in your neighbourhood have a kid or two, they still appear to be around. It's a bit like an optical illusion, because in terms of raw numbers there are still a lot of children about – especially in low- and middle-income countries, like India, where 26 per cent of the population is under 14, compared with 18 per cent in the US.[42] But it's fertility rates you have to look to when thinking about the future of such a population. In India, the fertility rate has been in freefall since 1960, reaching an all-time low in 2021, when it hit the below replacement-level of 2.0 for the first time (and it's sitting at 1.6 in Indian cities).[43]

A large population will still produce high numbers of people, even with a shrinking fertility rate, which is what we're seeing. But there aren't enough children being born to sustain ongoing growth or stave off demographic decline. In 1970, children under the age of 14 made up 37.5 per cent of the world's population. A mere half-century later, they make up only 26 per cent of the global population.[44] This number has nowhere to go but down, as you'll see in chapter 12: The end of having babies.

Why aren't there enough young people or kids being born? Because in the developed world we stopped working on the land, moved to cities, got rich, developed reliable birth control, and stopped having spare babies to till the fields and replace the ones that died in childhood. That leaves us with a novel conundrum, because we're used to having demographics that match what's called a population pyramid.

A population pyramid looks just like an Egyptian edifice, with a big chunky base, a smaller middle, and a pointy top.

The pointy top represents the old people, and in the classic model there aren't meant to be many of them, because life's hard and lots of things'll kill you. The chunky base represents the kids, and there are meant to be lots of them. But what's happening in many countries is that the pyramid is starting to look more like a rectangle. Eventually it could look like an inverted pyramid with a tiny pointy base of kids and big chunky top of geriatrics. What happens if you try and balance something like that? It's going to topple over.

Governments have noticed. They all have centres and institutes dedicated to the study of ageing populations and age-related diseases. In an interview with the journal *Nature Aging* in 2021, the geriatrician and director of America's National Institute on Aging (NIA) Luigi Ferrucci stated:

> What I can say to the people that are starting to study aging now is that I think we are on the verge of a revolution, that the science of aging is maturing and becoming a medical science, and a discipline that every doctor will have to be competent in to practice high-level medicine. There's never been a more appropriate moment to dedicate oneself to older adults. It has an important societal role, the emerging science is incredible and opportunities for new treatments are here. There is an aging population, the need is upon us.[45]

Governments are keen to keep us healthy for as long as possible, in no small part because it's much cheaper that way. As the gerontologist Nir Barzilai points out, a two-year increase in healthspan (the period we live unencumbered by chronic illness) could save \$7 trillion by 2050.[46] That's a huge saving for a very modest improvement to human health. Imagine how much more we could save if we radically extended human lifespans and healthspans and repurposed that money to promote human wellbeing on other fronts.

The counterfactual

Now imagine things keep on much as they are. The global population continues to grey, fewer children are born, economic growth slows and the global economy contracts. We can't afford pensions for everybody who needs them and worked their whole life expecting that safety net. Young generations whose economic prospects have been shattered by the global financial crisis, the coronavirus pandemic, rising automation, and humanity's great demographic shift, have no sympathy for the plight of the ageing cohorts whom they believe robbed them of their future. They don't feel this way because they are inherently unsympathetic, but because they are personally on the back foot, anxious and scared. There are more zero–sum games in a world of greater scarcity and this promotes tribalism. The young are angry with the old. The old never thought their twilight years would be like this. Everybody suffers.

Along with the risks posed by artificial intelligence, future pandemics and nuclear proliferation, this is another grenade that we're juggling precariously, as we sail towards that iceberg that is climate change. At present there is no indication that we've figured out how to change course. Governments around the world have tried strategy after strategy to get us to have more babies – including asking us if we would please consider doing it. This is tactically ludicrous. The right social, economic, cultural and technological preconditions are not in place for that to happen.

Some think the answer is for governments to try harder, by offering better financial incentives. Make childcare free, give big tax breaks to parents, and give them substantial wads of cash for taking the trouble to bring the next generation of humans into the world. I think that's an idea that's worth seriously considering – and there may be circumstances that arise over the next hundred years that render policies of that nature both necessary and desirable. I also think it would be presently smart to make early childcare and pre-primary education free in the countries that can afford to do so.

But my bigger bet, at present, is that even very generous incentives still won't be enough to *sustainably* boost fertility rates to above replacement levels in a majority of countries. We're barreling towards a very different kind of future and we have to start accepting that producing more people simply so that they can serve as cogs in a machine of economic growth is neither a plausible, nor desirable strategy for the future of humanity this century.

To begin to understand why, we need to take a look at our future economic prospects, and the impacts of automation on the bank accounts, sex lives, psychology and reproductive opportunities of the young. No leader is saying the obvious out loud: middle- and working-class people are being displaced; men and women are talking past each other and siloising; many young people do not have the psychological resilience, maturity or economic stability to marry and start families; and people are increasingly wondering what the point of life in such a world is. There's no bringing the old script of a stable job, spouse, picket fence and 2.3 children back. That dream is over. It's time to dare to dream bigger and pursue a future of radical longevity, intelligence enhancement, few deaths and few babies. That's a best-case scenario for humanity in the 21st century. But for many of us, this transition is going to hurt.

9

THE POST-WORK SOCIETY

The market will continue to throw millions of people out of the labor force as automation and technology improve. In order for society to continue to function and thrive when tens of millions of Americans don't have jobs, we will need to rethink the relationship between work and being able to pay for basic needs. And then, we will have to determine ways to convey the psychic and social benefits of work in other ways.

Andrew Yang, *The War on Normal People*

The goal of the future is full unemployment so we can play.

Arthur C Clarke, interview with Gene Youngblood

Imagine the life of a child born in Australia today. Let's call her Ava. Her life expectancy at birth is 85 years, but each year she lives it's likely to rise, with no fixed upper limit. By the time Ava's old enough to get a driver's licence, many cars will drive themselves. In fact, human drivers may even be outlawed in parts of the world by then. By the time she graduates from university, at least half the jobs in today's world will no longer exist, or will have been radically transformed.[1]

When teenage Ava goes out for burgers with friends, there's a good chance that the patty will be grown in a lab using stem cells. It could be 3D printed, cooked by a robot, and made of a customisable blend of exotic meats like lion and bear, all without inflicting a shred of harm on any animal. The lettuce and tomato on the burger could be grown in an energy-efficient, pesticide-free, vertical farm just a few blocks away, which would slash logistics costs and reduce the environmental impact of food production, rendering food cheaper, fresher and safer. The produce could even be nutritionally enriched using genetic engineering.

Ava will never have a part-time job operating a checkout in a supermarket, or flipping burgers at McDonalds. Her teen years won't involve that long-forgotten practice of hanging out at the mall. The mall will no longer exist in the form of physical storefronts on main streets, or in big shopping complexes. It will exist online, in the world of bits rather than atoms.

These days it's a truism to say that automation is transforming the workforce. We know that careers in logistics, retail and hospitality aren't smart bets for a kid growing up today – and it's worth thinking twice about a career in accounting, law, administration, human resources, dermatology or radiology. What we still haven't figured out is what the kids born today will grow up to become and how they will find fulfilment in life.

If government rhetoric about science, technology, engineering and mathematics (STEM) education is to be believed, they'll all grow up to become coding wizards putting those 'algorithms' on 'the blockchain' and building the future. Give them some robots to play with and they can be resilient digital natives. Yeah, except in twenty years the goalposts will have shifted and much more of the technical work we do today will be automated, further upping the intellectual and skill level ante required for humans to add value in the knowledge economy.

This should give us pause when we hear someone espousing the comforting narrative that automation historically creates as many jobs as it destroys. To determine the impacts of a productivity revolution on a society, you have to look beyond the number of jobs lost and the number of jobs created. It also matters what *types* of new jobs get created and who can do them.

In the 21st century, we're not replacing agricultural labour and muscle power with low-level cognitive tasks and repetitive manual tasks. We're replacing data-entry and factory, call centre, driving, retail and office jobs with those that disproportionately require high levels of education, intelligence, social skills and technical acumen. Jobs that many people, no matter how hard they try, or how much they study, will never be able to do.[2]

The robots are coming for your jobs

I have a great-aunt, Chris, who spent most of her working years as a neg-cutter in the Australian film industry. This used to be a trade: take all the film that was shot and splice bits together to make a coherent reel. One of the last films she worked on was Baz Luhrmann's *Moulin Rouge*, released in 2001. She found spots of work in later years at the National Archives in Canberra, but with digital technology revolutionising the industry, this profession is no more.

Her partner, my biological great-aunt, Maria, spent most of her career working on the phones at Singapore Airlines. This was back when people had to go to a travel agent, or call the airline to book or reschedule a flight. It wasn't a glamorous or high-paying career, but it was a steady job and she had great affection for the company for most of her time there. She was part of a community and was respected and well looked after.

Her experience was very different to the high-turnover situation in call centres today, in a labour market replete with casual and gig-economy jobs. Maria had a home at Singapore Airlines and she worked hard and appreciated that stability. Maria and Chris are now retired. They own their own home and are financially secure. But imagine what their future would look like if they were midway through their careers today and were suddenly made redundant. What alternatives would be available to two middle-aged women who didn't graduate from high school?

Like millions of others working in industries facing job cuts or obsolescence, my aunties would likely be left to compete for a shrinking pool of low-skilled, low-paid positions. Far from being rewarded for their decades in the workforce, they'd be forced to accept less security, be part of a high-turnover workplace community, and reap fewer of the associated benefits than they'd previously enjoyed.

Trying to secure their future would be doubly difficult in a world of high rents and even more unaffordable house prices. In their home city of Sydney, house prices are around thirteen times the median household income. An average millennial couple would have to work for eight years to save a deposit, while a single person would have to work for sixteen years – and that doesn't account for the fact that many young people are spending longer in higher education and part-time employment, and earn *less* than the median income for much of their twenties. Things were very different for the baby boomers, who entered the workforce younger. They could save a deposit in four years for a home that cost four times the average annual income.[3]

While Sydney is among the handful of most unaffordable cities in the world, property ownership, and the sense of stability it confers, is increasingly beyond the reach of young people and low-income earners in many cities across the developed world.[4] If they were starting out in life today, my aunties would find themselves embarking on a very different – and much less stable and satisfying – journey. Instead of buying in to society and believing that hard work will be rewarded, they'd be more likely to feel downtrodden and resent the systems and institutions that allow them to fall through the cracks.

A fraction of a bigger picture

The embrace of a post-work society is part of the larger story of our transhuman transition. Human societies are in the early stages of a great economic and social recalibration, as we head towards a future of greater productivity, generated by less human labour. If we are proactive in adapting our policies and institutions, this revolutionary period could present us with a suite of opportunities to reorient our lives around family, nature, learning, community, empathy and love in the short term, and the pursuit of a dynamically sustainable future in the long run.

These are the kinds of life priorities we should be raising young generations to focus on, instead of the frequently mindless, routine, and preposterously time-intensive exams, careers and credentialing exercises that many invest in for most waking hours of most days – all the while escaping into the social media vortex and dutifully swallowing their antidepressants to take the edge off the stress, exhaustion, loneliness and lack of higher purpose that is increasingly characteristic of modern life.

Unfortunately, the worst shocks and upheavals of this transition are yet to come. Lives will be turned upside-down and livelihoods stripped away by rapid automation before governments and communities know how to manage and mitigate these blows. The self-esteem and mental health of many will take sharp dives as their identity is forcibly uncoupled from the professions and responsibilities that once defined them.

Long before the Covid-19 pandemic struck – which has accelerated the pace of workplace automation, and will help shape the kinds of technical solutions businesses adopt from now on – it was widely reported that machines were coming for many of the tasks that make up a huge proportion of today's jobs. Susan Lund of the McKinsey Global Institute forecast that around the world, 'up to 375 million people may need to learn an entirely new occupation' by 2030.[5] In a 2017 McKinsey report, the authors also noted that 'about half of all work activities globally have the technical potential to be automated by adapting currently demonstrated technologies'.[6]

Of course, we didn't immediately automate all those tasks. Rates of automation are principally driven by cost-

effectiveness, along with institutional norms, social acceptance and the composition of the labour force, which varies
across countries. But that figure gives you some idea of what's
technically feasible now. So, how much more automation
could we see as new technologies emerge, evolve and get
cheaper?

The co-author of the seminal textbook on artificial
intelligence, Stuart Russell, thinks that 'as AI progresses, it is
certainly possible – perhaps even likely – that within the next
few decades essentially all routine physical and mental labor
will be done more cheaply by machines'.[7] If he's even in the
right ballpark, then the structure of human lives and economies
is going to change rapidly and radically this century.

Machines don't need to be able to do every aspect of a
job to take millions of jobs away – or in the case of a country
like China, which has the largest unskilled labour force in the
world, tens, or hundreds of millions of jobs.[8] Maybe a bot only
does half the tasks the person in the call centre used to do
and the customer is pinged back and forth. That still results
in something like a halving of employees if demand remains
constant. Trucks also don't need to be fully self-driving to carry
freight without a driver. They can travel in convoys, where the
first and last vehicle have a human driver as backup, while
those in the middle follow along autonomously.

In blue- and white-collar industries, the most at-risk jobs
are those that involve lots of routine and repetitive tasks –
whether it's order fulfilment, call centre jobs, factory work,
legal discovery and conveyancing, or radiology. Lots of aspects
of human resources, secretarial work and teaching will also
likely be automated too – and blended learning will change the
nature of teaching roles significantly at all levels of education

(though it would take another book to explore the promise and pitfalls of various models of educational reform). The most resilient jobs in the coming years and decades will be those that are non-routine, disproportionately require higher intelligence and social skills, and for which there is increasing demand. Think cleaning, nursing and aged care, and knowledge generation, or creative work, in the sciences, technology, engineering, and the arts.[9]

Yet I don't think those distinctions get to the heart of what we're facing as a society in the age of automation. Having a job isn't synonymous with stability, or a good life. And there's a world of difference between being sure you can get work as a cleaner, and being sure you can get work as an AI developer. Both skillsets will probably render you very employable for a long time to come. But only one is likely to provide you with financial security, social status, and the social, sexual and psychological benefits that status and stability confer.

A growing divide

'So, what do you do?' is still one of the first questions most of us utter when we make a new acquaintance. Work is a huge part of most people's identity, and, for men in particular, the ability to earn money and garner social status through professional achievements is closely tied to their sexual and reproductive success. In a world where more men are unable to attain the economic stability necessary to attract a partner, what becomes of social cohesion, the relationship between the sexes, family structures, and mental health and happiness?

It's already happening. As the American tech entrepreneur and politician Andrew Yang points out in his 2018
book *The War on Normal People*, five million manufacturing
jobs have evaporated in America since 2000, four million of
which were the result of automation. Most of the workers
affected were men, and 'instead of finding new jobs, a lot of
those people left the workforce and didn't come back'. The
historically low labour force participation rate in the US and
the decline of manufacturing jobs also coincides with other
worrying trends, such as the recent opioid epidemic, and
the tipping point around 2017 when overdoses and suicides
overtook car accidents as leading causes of preventable
death.[10]

Men who are not working, have low social status, and lack
stable sexual partnerships tend to engage in more anti-social
behaviour. As Yang reports, 'high rates of unemployment and
underemployment are linked to an array of social problems,
including substance abuse, domestic violence, child abuse,
and depression'.[11] In the US in 2016, '22 percent of men
between the ages of 21 and 30 with less than a bachelor's
degree reported not working at all in the previous year – up
from only 9.5 percent in 2000'.[12] More men are dropping out
of the labour force and 'young men without college degrees
have replaced 75 percent of the time they used to spend
working with time on the computer, mostly playing video
games'.[13]

Of course, there's nothing inherently bad about enjoying
video games. They're a technological marvel that provide
many people (young men in particular) with things that are
increasingly difficult to obtain in real life: immersive, exciting
adventures, strategising, going to war, quashing rebellions,

banding together with a tribe, and rescuing pretty women. Rewards are built in to the virtual landscape and are attainable in proportion to effort and skill – unlike in the workforce, where you can be the best darn barista or 'sandwich artist' in the world, and still earn minimum wage.

I'm not afraid of a future where humans spend more of our waking hours in virtual worlds. As you're about to see, I think that will be a necessary step later this century. What's worrying is the asymmetry of the uptake of these technologies and the social instability that could follow, as early adopters are driven into alternate realities by a lack of options in what most of us currently revere as the 'real' world. If those drifters disproportionately remain male, sex ratios in the dating pool will continue to skew, and eligible men will become even more scarce, bestowing enormous sexual power on the minority of men who are still considered sexually viable by women. This is not a recipe for a flourishing human society.

We need to talk about men

We're very uncomfortable talking about this in Western societies, but men are, increasingly, a problem. Not in the 'privileged beasts of the patriarchy' kind of way. If anything, in something close to the opposite of that. Men are falling in our world harder and faster than women and working-class men are having a particularly tough time of things. In blue-collar industries in the US, automation has affected men more than women. In middle-skill jobs, it's affected women more – but women have historically been more resilient, successfully upgrading their skills and finding better paid positions. Men in middle-skill occupations have, by contrast, settled more frequently for lower skilled and lower paid work.[14] There are two key reasons for this. Average sex differences, and diverging education levels.

On average, women have superior verbal skills and social and emotional intelligence.[15] In the short term, these are strong advantages in the modern workforce, as human abilities in these domains are far superior to those of machines. Male superiority manifests in stronger visuospatial skills, as well as superior mathematical ability at the very highest levels of maths. Of course there's wide variation among the individuals of each sex, but on average, the adage that women are more interested in people and men are more interested in things, holds true.

At a glance it might seem like both sexes can leverage their relative advantages reasonably well. But look closer. The male-dominated industries that are booming, like the physical sciences, technology and engineering have high barriers to entry. Those booms are great news for a minority of highly intelligent men, but they're of scant use to average or below-average men. The female-dominated industries that are booming, like aged care, childcare and healthcare, are more diverse and present more opportunities for sociable women along a broader stretch of the bell curve to reskill – and they're disproportionately seizing those opportunities.

Many readers will be surprised to hear this, but women are also out-graduating their male counterparts at a phenomenal rate. The US is on track to churn out 47 per cent more female college graduates than males – or three women for every two men – in 2023.[16] According to the demographer Albert Esteve and colleagues, 'in 2010, the proportion of women aged 25–29 with a college education was higher than that of men in more than 139 countries that altogether represent 86 percent of the world's population'. By 2050, it's expected that 'women will have more education than men in nearly every country in the world'.[17]

As women rise in the workforce and significantly out-graduate their male counterparts, they look for partners who have equal or greater education levels, status and earning potential – and who are acceptably physically attractive. Where are they? They don't exist any more. Or rather, there simply aren't enough of them to go around. This skews the dynamics of the sexual marketplace profoundly.

Women simply don't want to pair up with unemployed or low-status men. When we see a man on some version of struggle street, looking dishevelled, shoulders hunched over submissively, wringing his hands anxiously, and talking about how he's kind of a homebody and spends a lot of time playing video games, our brains scream 'no way, get me outta here'.

The men who still meet female criteria for sex, dating and mating are in exceptionally high demand. The men who don't are increasingly frozen out of the game of life altogether. This is creating a perverse dynamic in the sexual and romantic lives of young people. We're increasingly talking past each other, failing to pair up, and channelling a lot of that pent-up frustration into fourth-wave feminism, men's rights activism, and any movement that seeks to overhaul an establishment we feel is failing us. We'll explore this dynamic further in the next chapter.

For now, suffice to say we're living in a world where a life-script based around the steady 9–5 is starting to crumble. A majority of humans will not be able to compete with machines as producers of economic value by the century's end (and probably long before). But humanity *will* be able to reap incredible rewards from the massive dividends machines unlock. AI alone will add tens of trillions of dollars to the size of the global economy and will drive the cost of living down, as the high cost of human wages is priced into goods and services. Once we decouple work from survival, we'll have the freedom to build our lives around more meaningful things that involve leisure, learning and pleasure. But as promising as that sounds, it matters in what order the dominoes fall.

Sympathy for the devil (I mean, the politicians)

Imagine you're the leader of a rich, powerful country. What would you do if you knew with moderate to high certainty that millions of people in your nation were working towards goals and pursuing life-scripts that would increasingly become unattainable? For the most part your citizens are utterly unprepared and walking blindly towards the edge of a cliff. Would you warn them?

On the one hand, knowledge is power. On the other hand, a little learning is a dangerous thing. I used to be incredibly critical of politicians who failed to honestly answer questions about the future job prospects of the young and middle-aged. But I've been forcing myself to think about what I'd do if I were a leader today, and it's really tough.

Politicians tend to be very focused on getting people back to work and keeping them working. The anthropologist Dave Graeber, who penned the popular bestseller *Bullshit Jobs*, wrote – with bit of a conspiratorial gloss – that there are so many dull, paper-pushing office jobs that feel pointless to those who perform them because 'the ruling class has figured out that a happy and productive population with free time on their hands is a mortal danger'.[18] The thing is, to the extent our leaders believe this, they're right. Suddenly unleashing people from their cubicles *en masse* is dangerous.

It's untenable in the short term, because a lot of dull, routine jobs are still essential to the functioning of society. But it's *dangerous* because we haven't fleshed out the next chapter of the human social narrative, or put cushioning infrastructure in place. It sounds great to say, 'give people a universal basic income (a guaranteed wage paid automatically by the government) and let them decide what to do with their lives'. The libertarian in me is on board with that. Until I remember that almost half the human population has an IQ below 100. And that even many very smart people lack the personality traits that drive them towards independence, freethinking and the kind of creativity that thrives when nobody is telling you where to go and what to do.

In 1970, the American engineer, architect and futurist Buckminster Fuller told *New York Magazine*:

> We must do away with the absolutely specious notion that everybody has to earn a living. It is a fact today that one in ten thousand of us can make a technological breakthrough capable of supporting all the rest. The youth of today are absolutely right in recognizing this nonsense of earning a living. We keep inventing jobs because of this false idea that everybody has to be employed at some kind of drudgery because, according to Malthusian-Darwinian theory, he must justify his

right to exist. So we have inspectors of inspectors and people making instruments for inspectors to inspect inspectors. The true business of people should be to go back to school and think about whatever it was they were thinking about before somebody came along and told them they had to earn a living.[19]

I love this idea, but Bucky missed a crucial point. Not all humans were designed like Buckminster Fuller, in the form of oddball, big-picture thinkers. Some people seem much happier functioning as the human equivalent of worker bees, soldier ants and breeders (I don't mean that to sound derisive, I say it because it really appears to be true). Not exclusively, of course. Everybody has rich and varied dimensions to their being and nobody wants to be trapped in a life where they're forced to do dehumanising work, or to work so much that life becomes unbalanced and one-dimensional. But take away work entirely and you take away humanity's last remaining church – the kind of coercive, but also kind of cooperative, thing that keeps us all in check and makes it clear what we're meant to be doing with our time. It's a quasi-religion that most people still believe in, and for all its faults, it's one of the few things (having children is the other big one) that brings structure and purpose to most people's lives.

It's difficult to restructure societies in a timely fashion when the old life-scripts are falling unfathomably fast. We're yet to formulate the right safety nets that will allow people to live with dignity, while creating new avenues for those displaced from the workforce (which might eventually be most or all of us) to find new kinds of satisfying life-scripts.

As is so often the case in times of turmoil, I suspect the richest countries, with longstanding social welfare programs and the means to rapidly scale them up, will have the smoothest ride – but even then, I think it will be a bumpy one.

The double-edged sword of a universal basic income

My bet is that we're going to see lots of governments trial universal basic incomes (UBI) in the 2020s. That means they're going to give all but the very richest among us money for nothing. This will be a necessary and positive step in human social evolution. But I also worry that the tide's turning too fast. UBI proposals have gone from untenable, in the eyes of mainstream politicians, to sexy overnight. When a pandemic forces you to hand out eye-watering sums of money, using the infrastructure of presently inefficient welfare systems, a UBI starts to sound like a more streamlined and palatable option.

The point of a basic income is that it redistributes the spoils of automation and allows everybody in a society to have their basic needs met and to live with dignity. The hope is that it will also encourage risk-taking, creativity and innovation. A basic income can alleviate the scarcity mindset and short-term thinking that poverty breeds and give us the freedom to take time out and upskill, reskill, start businesses, think, create, and spend more time with our families.

This would be wonderful in the hands of people who currently have social anchors and community ties, a sense of purpose, and some ambition. But in the hands of those who have already given up, have few (if any) social connections, and are on the cusp of bottoming out of society, it's more likely to be a disaster.

This is absolutely the right decade to start trialling basic income experiments. But if we do start rolling out these programs state- and nation-wide, I don't think we should give people large sums of money *completely* unconditionally. First, we have to cultivate new social values and develop future-oriented institutions, education systems and life-scripts. I'm not suggesting we should encourage governments to create convoluted and coercive social welfare programs instead, or fail to reform those they already have. But there is room to consider a transitional middle ground.

The economist Daniel Susskind is of a similar mind. He proposed the idea of a conditional basic income (CBI) in his 2020 book *A World Without Work*. My hunch is that a CBI might be a good way to start. Actually, I think it would be best to run many experiments in parallel and see how different communities (and sections of the same community) perform with different welfare incentives. But if I could play God and design a welfare system for the community I live in, here are my current thoughts about how I'd do it.

There's one key condition I think we should attach to our hand-outs for the 2020s if we start scaling them up to close to a living wage. If you're not working or studying for more than 15 hours a week, give or take, you spend one day (or part thereof) doing something useful or meaningful – like participating in organised efforts to clean up local parks and beaches, working at a soup kitchen, or visiting elderly members of your community. Like going to church on a Sunday, the idea is to bring back a shred of the community-mindedness, social connection, and sense of purpose and responsibility that is falling away in modern societies. For some, this might be the only meaningful interaction they have with the outside world, or people outside their family unit and social media echo chamber.

A CBI policy could also target young people with aptitude and train them for essential roles – on an entirely opt-in basis, of course. Two of the biggest areas where contributions could be made are in childcare and aged care. Teenagers and young adults could start in an interning role, with supervision and background checks, but no formal qualifications. They could learn on the job, and if they enjoy the contributions they're making, they'd have the option to take on additional hours, increase their payment, and work for accreditation. They should be well incentivised to do this, through attractive rates of remuneration and career development opportunities.

Instead of taking away the basic welfare safety net when people start working more, as many countries do, we should make that safety net guaranteed. Then give people *more* money to reward them for working in essential roles. In the short term we're at risk of facing a skills shortage in aged care, childcare, nursing and parts of medicine. If young people who have a natural aptitude for these roles grow up believing that work's bullshit, and there's no point studying or aspiring to a career because many of their peers aren't, our societies and economies (which would be sustaining huge social welfare programs at this point) will be in trouble.

A CBI policy would be a great way of getting more people into what are currently low-paid, high-turnover industries. And with millions more adults who aren't on a career track popping into an aged care facility for a few hours once a week, and having the freedom to do their own thing for the rest of their time, they would not be coming in with the mindset of an underpaid, oppressed, poor person living paycheck to paycheck. They would be more likely to come in, if they've chosen that industry, with a genuine urge to help and connect.

These kinds of caring roles won't be for everyone, and we should create enormous flexibility so that all CBI recipients can choose something that feels worthwhile and meaningful *to them* – and that's within the realm of their natural abilities. It's a bit like national service, but you don't have to have anything to do with the military, you're not forced into a rigid one-year program, and it's not about national pride. Of course, if getting free money on the condition that you tutor some refugee kids for a few hours each week, coach a sports team, mow someone's lawn, or do some shopping and read to an elderly person sounds *that* horrifying, then good news, you can forego your CBI and earn your own money instead.

I'm aware that a CBI policy, like the one I've described, may prove too costly to administer and implement. It might be more expedient to just pay people more to work in essential industries, in addition to a standard UBI, which is unconditional. I'm also sensitive to concerns about empowering governments to compel us to behave in a certain way that they deem prosocial, for 'the greater good'. The spirit in which I'm exploring a CBI is not to state with total confidence that I think it's the best policy. It's to foreground the need to have some kind of system that keeps communities together and humans interacting face to face in these transitional times, while we are still human. As there are so few proposals for how to do this, we have to start somewhere.

What you need is a gram of soma

This is not where the story ends, though. We're going to need more than occasional face time and community service to get us through the next few decades and beyond. For years, I've been telling anyone who'll listen that the next step is to invent soma. That's the name of the feel-good drug that citizens take in Aldous Huxley's novel *Brave New World*, boasting, 'all the advantages of Christianity and alcohol; none of their defects'.

The standard reading of soma in *Brave New World* is as an agent of pacification in a dystopian society. Independent thought and the highs and lows of human experience are levelled out by advanced psychopharmacology and crucial aspects of our humanity are blunted, or lost. This is not what I have in mind. My vision of soma is a fuzzy one – an idea that reaches towards the concept of building more fulfilling digital lives, and changing brain states and moods in far more precise ways than routinely handing out something like Prozac.

We already have soma-esque products in our world, like social media. But these platforms weren't designed to keep us virtually engaged in a healthy or fulfilling fashion. They're data-harvesting farms and advertising delivery drips, not human enrichment platforms. There's abundant research linking social media use to anxiety, depression and low self-esteem.[20] That's not the kind of soma we want people consuming *en masse*. Nor are today's antidepressants, which come laden with side effects – and we're still not sure why they work, under what circumstances, to what extent, and if they should be taken long term.

We need better drugs and better virtual worlds, and in all likelihood, we'll invent them. But why are we going to need them? Well, what's the alternative in a world where the average person can't earn enough to live with dignity and is reliant on basic income payments? How do they fill their days? What else is going to confer a sense of progress and achievement in life? Anyone who's suggesting we just magically 'create' more jobs and put people to work hasn't come to terms with how scant and poorly remunerated the tasks left for people with average or below average intelligence will be.

The only alternative I can think of during this transitional period for humanity is to encourage people to have more children and to spend their days investing in family and community. But I don't think that strategy will work, as you'll see in the next few chapters. Marriage, long-term pair-bonding, and procreation are in steep decline (and they're in steepest decline among low-income earners in the West). In a basic-income world, with more advanced artificial intelligence, who's going to sign up for the astronomical costs and responsibilities of raising children – most of whom won't grow up to have jobs, or pursue the familiar human life-scripts we expect when we have children today? It's a hard sell against the alternative of a compelling digital universe, where new kinds of experiences and pleasures – like virtual sex and perfect partnerships with AI beings – are available on tap. More on this in the final chapters.

We're tarrying in acknowledging this, but I think the tide has already turned on 'real life' for a growing cohort of humans. Reality no longer offers satisfying rewards or regular feel-good milestones of progress, the way short-burst dopamine hits in the form of 'shares' and 'likes' do.

Low-status, low-skilled, low-IQ and unemployed people – especially those who are single and don't have children – have the weakest buy-in to reality, and the fewest opportunities to make their way in this brave new world.

As sad as that is, pretending we can make it better by winding the clock back to something resembling the 20th century is a fantasy. We need to look ahead and plan for a new reality that can see us through these transition times humanely. But we need to be honest about the fact that some people *are* going to suffer in the short term. No major social transition occurs without pain, and no period in human history has been devoid of suffering. Our task is not to bury our heads in the sand, it's to look at where the pain is likely to be felt most acutely, and do everything in our power to minimise it, while creating the necessary preconditions for a sustainable future that offers radical abundance to all.

The picture is always bigger

I know I've barely touched on the risks of somatisation, or how it could render a herd-like populace vulnerable to control by bad actors with nefarious intentions. But I'm telling a story that connects the dots between key phenomena in our transhuman world from a zoomed-out perspective. I have to pick and choose what talking points to home in on to keep the narrative moving forward. Rest assured, I know there are risks associated with anything resembling soma, just as there are risks associated with maintaining the status quo. I've chosen to broadly focus on how soma could function and what role it could play in our societies, because I think it's likely we're going to invent it as an adjacent possibility, even if we don't mean to.

I also think we are going to *need* soma as automation ramps up. People are very reluctant to have an honest conversation about the fact that we're going to see an ever-sharper divide between the world's haves and have-nots in the coming years, unless something changes soon – though I think a merger with AI will follow as the great leveller. We're comfortable talking about the fact that the economic returns from labour are being outflanked by returns from capital, leading those who already own homes and shares to get richer, while those who work as wage slaves get poorer.

But we're not comfortable stating the other obvious fact. The most adaptive kind of human (that is, the best placed to survive and thrive) in this novel transhuman world is a smart human. Many people will deny that there's such a thing as intelligence differences – there are just different kinds of intelligence and they're all equally valuable. But all types of intelligence tend to be strongly correlated – we just notice the exceptions, in the form of brilliant thinkers with terrible social skills, or clever wordsmiths who can't read maps, because they stand out.

To the extent our leaders are honest about the demands of our novel world, we hear calls to 'educate' people more. It's almost invariably unclear what this means. We hear slogans about fostering 'creative and critical thinking' and preparing kids to be 'future ready'. But in practice this seems to amount to little more than doing some project-based learning in school, continuing to funnel ever more young people into the expensive Ponzi scheme of higher education, and priming them to believe in the pursuit of a crumbling 20th-century life-script.

My argument here is that there's no resurrecting that life-script for a majority of humans, no matter how much we might wish to. The rapid onslaught of automation and human displacement is inexorable. This brings us back to the 'two worlds problem'. In the coming years and decades, we're going to face a dual crisis of economic displacement and of purpose. As we ride out this transitional period, we'll need to get better at 'humaning' in the short term and prioritise it more in our societies. That means more emphasis on family, community and face time, and attributing greater value to kindness, prosocial behaviour and investment in others.

We'll need to destigmatise under- and unemployment too, while still encouraging people to actively participate in society in other ways. But we must also face the fact that the age of relearning how to human will be a temporary transitional phase for a species that's going to out-evolve itself and become something other than human. Perhaps not long after we finally figure out how to human well.

The urgent task of refocusing our societal values on human connection over credentialism and earning to live will be hard to pull off in the 2020s. It's probably too late for many of today's kids, who are routinely guzzling brain-crack from their addictive, ever-vibrating devices, which many have been attached to since infancy. As you'll see in the next chapter, young people today are doing it tough, growing up in a world where the old life-scripts are evaporating underfoot. I fear they will be a major part of the collateral damage of our transhuman transition. But they're also curious harbingers of new, and hopefully much better, things to come.

10

A GENERATION OF KIDULTS

Our evolved biology can interact with culture and economics to generate misery for millions of people.

Rob Brooks, *Sex, Genes & Rock 'n' Roll*

All animals are equal, but some animals are more equal than others.

George Orwell, *Animal Farm*

It's time to talk about the generations whose sex lives, economic prospects, values and lifestyles could change the future. This is a crucial juncture in the story of our transhuman transition, where multiple threads converge. While life-extension technologies and the emergence of a post-work society could usher in a lasting era of morphological freedom and abundance, I fear a large proportion of today's young people are going to suffer as the collateral damage of our transition, floundering in the no-man's land between the human and posthuman worlds.

We'll spend most of our time looking at Gen Z, who are today's 10- to 27-year-olds. As we'll see, this cohort is less equipped than any previous generation to start pair-bonding and giving birth to the humans of tomorrow. Characterised by arrested development, social lives that increasingly take place online, fragility, risk aversion, uncertain economic prospects paired with expectations of upwards social mobility, decreased sexual activity, and the growing belief that sex and gender are fluid, they are stepping stones on the pathway to a world where having babies is on its way out, and life extension is on its way in, where work is on its way down, and automation and AI are on their way up.[1] This is not where the story of our transhuman transition ends, but it's an interesting place to take a screenshot.

The new kids on the block

We're used to thinking of kids these days as Millennials. But the American psychologist Jean Twenge argues that there was a generational break in 1995, when a new cohort of digital natives was born and grew up in the age of smartphones and social media. Twenge calls this group iGen or the internet generation, as they 'grew up with cell phones, had an Instagram page before they started high school, and do not remember a time before the Internet'. What makes iGen, or Gen Z, different from previous generations? You could sum it up with the phrase: *arrested development*.

Twenge's research finds that Gen Z teens are less likely to go out without their parents than previous generations at the same age. They're also less likely, even as teenagers, to be left alone unsupervised after school, or to walk to parks and public places as children without supervision. Other activities that once conferred responsibility and independence in adolescence, like learning to drive and working for pay, are being eschewed in record numbers too. Instead of receiving an allowance from their parents to manage independently, Gen Z teens more frequently appeal to their parents for things they want or need.

Then there's social life and dating. Gen Z are putting off trying alcohol until later in adolescence and spend less time partying than previous generations. The number of US teens who meet up with friends every day has halved in fifteen years and 'time spent with friends in person has been replaced by time spent with friends (and virtual friends) online'. Gen Z are also much less likely to date, have sex, get married, or become pregnant while still in high school – and they're much more likely to be virgins as young adults. Twenge reports that 'more than twice as many iGen'ers and late Millennials (those born in the 1990s) in their early twenties (16%) had not had sex at all since age 18 compared to GenX'ers at the same age (6%)'.

Lots of those trends sound really positive. Alcohol and developing brains don't mix well, and becoming a mother at 17 is probably not an ideal life choice for most young women. Unfortunately, it's not all good news. Gen Z teens are spending more time online than Millennials at the same age and they're suffering from the worst youth mental health crisis on record. Teens have reported increasing satisfaction with their lives as a whole since the 1980s. Millennials were optimistic teens, known, in Twenge's account, for 'their more positive self-views, higher narcissism, and heightened aspirations compared to previous generations'. But when Gen Z emerged on the scene, everything went backwards, fast. This plummeting in teen life satisfaction coincides perfectly with the rise of smartphones and social media.

Rates of teens in America who report feeling like they 'can't do anything right', that their 'life is not useful', and that they 'do not enjoy life' have skyrocketed since 2012 – a trend that is consistent across 8th, 10th and 12th grade cohorts. Between 2012–2015, rates of teen boys reporting depression increased by 21 per cent. For girls, who use social media more frequently, the increase was 50 per cent over the same period. Similar trends have been found in surveys of freshman college students.

Sharp rises have also been seen in clinically diagnosable episodes of major depression, with one in five teen girls having experienced a major depressive episode by 2015. Major depression is a significant risk factor for suicide, and we're seeing the uptick in teen depression translate into more suicide attempts and completed suicides among the young of both sexes.[2] This gives us some reason to doubt the hypothesis that rising depression among teens is simply due to increased reporting and decreased stigma surrounding mental health issues. Even more alarming is the fact that 'two and a half times more 12- to 14-year-olds killed themselves in America in 2015, versus 2007'.[3]

Although Twenge's research focuses on Gen Z'ers in the US, she notes:

> Researchers around the world are documenting many of the same trends, with new studies constantly appearing. The Internet and smartphone boom hit other industrialized countries at about the same time as these technologies took hold in the United States, and the consequences are likely to be similar.[4]

For better or worse (and it seems that it's both for better *and* worse) kids these days are increasingly risk averse, and taking longer to undergo most rites of passage into adulthood. They are the clear manifestation of a *slow life strategy*, which humans adopt in times of abundance. When things are good, we have fewer children and invest more resources in each one – think plenty of calories, preschool, tutoring, extra-curricular activities, and the sacrifices that are made to get them through higher education. In America and other developed countries, Gen Z is on track to be the most educated generation in history – 57 per cent of American Gen Z'ers over 18 were enrolled in college in 2018.[5] That sounds great, but they're also the most coddled and over-parented generation, and have had very little responsibility to shoulder in their lives to date. Perhaps not surprisingly, many have arrived on university campuses and in modern workplaces with the mentality that everywhere in life should be a 'safe space', where nothing uncomfortable, complex or awkward ever happens.

In Twenge's words, 'instead of resenting being treated like children, iGen'ers wish they could stay children for longer'. She reports that Gen Z teens score higher than previous generations on assessments of their maturity fears. They're more likely to agree with statements like 'I wish that I could return to the security of childhood' and 'The happiest time in life is when you were a child'. And they're less likely to agree with statements like 'I would rather be an adult than a child' and 'I feel happy that I am not a child any more'.

The stretching out of the circle of life

In the 19th century, there were children and adults. In the 20th century, there were children, teenagers and adults. In the 21st century there are children, teenagers, kidults and adults. The maturation process is being stretched out to an unprecedented degree, in a way that would befit a world where a lifespan of over 100 was the norm, where we remained vital and healthy well into what is now considered old age, and where our fertility windows were no longer so constricted. But the harsh reality for Millennials and Gen Z, who are today's 10- to 41-year-olds, is that they've been pushed in this direction of extended maturation by the economic demands of a knowledge economy *before* all the other stuff has arrived to make this a sane or fulfilling life arc.

As the world's current and emerging crop of peak fertile humans, we're in a unique conundrum, which successive generations will probably be lucky enough to dodge. The stage between the ages of roughly 15 and 40 is arguably the most important period in a person's life. It's not necessarily the most special or the most fulfilling time, but it's the most important because those were the crucial years where we did the things that we are literally programmed to live for in the ancestral environment. Alongside surviving, we pair-bonded, we had children, and we helped them to survive with the help of our kin-ordered tribes. Rates of violence and infant mortality were high, but we had communities, and we had a clear social script and an unwavering sense of purpose.

Today, this narrative has been inverted. Rates of violence and infant mortality are at record lows. Yet the established order and life-scripts of old are crumbling, and we are in the thick of the two-worlds problem. We're still programmed to pursue our deeply hardwired evolutionary goals of sex, pair-bonding and mating. Yet we're living in a modern culture that diverts our time and energy *away* from those pursuits, which are not valued by the economy, and are often difficult to secure and maintain. The crunch is particularly difficult for young women, who now routinely spend their best childbearing years studying, or working to get a foot on the first rung of a career ladder. Economic success is also becoming ever more of a winner-takes-all game, which is skewing sexual success for males, and romantic success for females, into a losing proposition for most, as we're about to see.

But this is a complex story with many facets. No generation has a single, homogenous identity. Plenty of Gen Z'ers aren't fragile snowflakes, some of them are having lots of sex, and many are in happy, fulfilling relationships. Nevertheless, the story that matters from the point of view of our trans-human transition is the overall change in attitudes, behaviours and available life-scripts for young people. And the life-script that *really* matters when it comes to determining the future of humanity is the readiness of our next generation to pair up and raise the humans of tomorrow, combined with their willingness to invest in *human* identities, circumscribed by the limits of human biology. Here's the backstory for why Gen Z is set to struggle, relative to previous generations.

When was the last time you got laid?

Young people have never put so much time and effort into looking fuckable on social media while having such paltry amounts of sex. According to General Social Survey data, the rates of Americans aged 18–29 who had not had sex in a 12-month period more than doubled in the decade since 2008. In 2018, nearly a quarter of young Americans had not had sex in the past year. This decline in sexual activity was particularly marked among young men, whose rates of sexlessness in a given year nearly trebled in a decade, rising from 10 per cent in 2008, to *28 per cent* at the end of the ten-year period.[6] More than a quarter of young American men having no sex at all for a year. If that figure doesn't worry you, it should.

An even starker trend of a declining interest in sex, or a growing inability to obtain it, is apparent in Japan, where nearly a quarter of adults aged 18–39 are virgins. Interestingly, the rates of male and female virginity in Japan are almost exactly the same. As Eric Mack reported in *Forbes* in 2019, the rates of those remaining virgins into their 30s in Japan are also higher than in other developed countries like the US, the UK and Australia. Notably, 'men in the lowest income brackets were also 10 to 20 times more likely to be virgins than those that earned the most'.[7]

Several journalists and academics have suggested, based on data collected by apps like Tinder, Hinge and OKCupid, that the modern dating market roughly follows a Pareto distribution (or the 80/20 rule). Based on who women are 'liking' or 'swiping right' on, it appears that close to 80 per cent of women are only interested in the top 20 per cent of men. That leaves the other 80 per cent of men to vie for the bottom 20 per cent of women – which they understandably think of as slim pickings.[8]

Throughout human history, men have disproportionately been 'evolutionary losers'. Every fertile womb is valuable, but not every seed is. One man can impregnate thousands of women, as Genghis Khan did, condemning hundreds of his male counterparts (those he hadn't already killed) to become evolutionary dead-ends. But we have cultivated more egalitarian cultural ideals in our wealthy modern societies, and we have high expectations of what a good life looks like for us as individuals, which includes having access to desirable mates for sex, relationships and procreation. Most people want the white picket fence (or the knowledge they could obtain it if they chose) and the ego boost and sense of wholeness that comes with being desired.

Globalising the dating pool and placing an enormous smorgasbord of options in swiping distance has elevated the expectations of many, who no longer want to settle for average Joe up the road, when there is the possibility (however unlikely) of meeting Spectacularly Successful Simon. And why settle for Sensible Suzy, when you could have Sandra, Stephanie, *and* Suzy? That is, if you're lucky enough to be among the minority of desirable men who could achieve such an outcome.

While success is amplified for the most attractive and high-status people on dating apps, who rapidly select each other out of the dating pool (or who take themselves out of the relationship market by playing the field), failure compounds at the other end of the spectrum and is felt on a much larger scale. The psychological distress for the male have-nots is also likely higher than in the past, because the sheer number of rejections they receive is made starkly apparent to app users, who are able to see the thousands of possible matches that are passing them by. A young man today can experience more rejection in a day than he'd have experienced in an entire hunter-gatherer lifetime.

Women are often rejected at a later stage in the game, when they make a bid for monogamy and investment. We don't talk about their pain and distress as much as we should. I think it really is analogous to what a lot of male incels (men who are involuntarily celibate, or believe they are so unattractive as to be incapable of finding mates) suffer – the thwarting of a hardwired desire to pair-bond, nurture and have children with a loving, invested partner. But almost all young women at least have the privilege of being sexually desired. That might not make them happy, but it's still a less dehumanising state than being sexually invisible. As the possessors of wombs, most women are free to have children if they choose to. We are not frozen out of the game of life entirely, as sexually invisible men are. Our white-picket-fence prospects, however, are getting grimmer.

When desirable males are scarce, a minority of men control the dating and mating market. We don't just see this in humans by the way, but in all monogamous, or monogam*ish*, species. As the journalist Jon Birger reports in his book, *Date-onomics: How dating became a lopsided numbers game*, male fidelity varies wildly in pond cichlids (a type of fish) when you start playing with sex ratios. When an even sex ratio of 6:6 was skewed so that males outnumbered females 7:5, male desertion rates halved 'from 22 percent to

11 percent'. Females were also choosier about their mates, males fought and competed for sexual access, the successful males carefully guarded and protected their females and were more invested in parenting. When the sex ratio was skewed the other way, with females outnumbering males 7:5, 'male desertion rates more than doubled – from 22 percent to 51 percent' and levels of male parental investment declined.[9]

Birger also cites the psychologists Paul F Secord and Marcia Guttentag, who explored historical literature and data from many societies including ancient Greece and medieval Europe. They found that societies valued romance and courtship more when females were in short supply. When women were in oversupply, they were more likely to be valued as sex objects and treated more disposably. In their words: 'The outstanding characteristic of times when women were in oversupply would be that men would *not* remain committed to the same woman throughout her childbearing years'.[10]

Granted, women are not literally in oversupply in Western countries. But they are oversupplied relative to male counterparts of equal or higher socioeconomic status. As such, the functional dynamic for women in the dating pool is one of female abundance and male scarcity, as a majority of women compete for a minority of sexually visible and desirable men. The result? Accelerating decline in marriage and fertility rates, a boom in men's rights activism and new waves of feminism (not coincidentally among those in their peak reproductive years), and a lot of miserable young people who are being thwarted in the pursuit of the very thing they're hardwired to pursue above all else, bar survival: the passing on of their genes. It's an oversimplification but I like to think of it as an epidemic of sexless men and loveless women.

When enough men and women start talking past each other and failing to pair up, it starts to seem more palatable, or even necessary, to forge stronger alliances with members of your own sex. For the most disgruntled men, that's led to a growth in movements like Men Going Their Own Way (MGTOW) where they swear off women for being too demanding and not worth the investment – the fear for average and below-average men is that we'll harvest their resources, cheat on them and extract child support payments for the next 18 years for kids that may or may not be theirs. While that should hardly be your dominant expectation for what most women will do, it's not a totally unfounded fear and some men *are* at higher risk of facing that reality than others.

On the flipside, I think the scarcity of marriageable men is fuelling a lot of the modern push for women to be self-reliant #BossWomen. Most women have no real desire to live a swinging-dick, cutthroat corporate lifestyle. But almost all women want kids. Kids are expensive and if you're going to go it alone, as an increasing number of women in their 30s look poised to, you need money. Men no longer look like a sound bet for the life-plan many women desire. Without a high probability of finding a solid male partner, we've begun (at least subconsciously) entertaining the notion that we'll be better off if we trample men as hard as we can economically and hoard those resources for ourselves. But that sounds icky, so we package it under the brand of *female empowerment.*

Nobody's vocalising the link, but I think a lot of women are intuiting that there's no Prince Charming coming to sweep them off their feet and that they need to make their own way in the world. Hence the sudden proliferation of clickbait, advertising and documentaries exploring the question of whether women should routinely freeze their eggs so that they can have careers and children on a less compressed timeline, regardless of whether they find a man. If that's the bind you find yourself in, I get why you'd also gravitate to *post hoc* rationalisations about how empowered you are as a strong, single, independent career woman. But I think this confession by the writer Briony Smith is closer to the truth. Aged 32, she wrote:

Loneliness is physical

It's a dull sort of pain, like a poke in the eye or the slow ebb of cramps. Often I don't feel it for a while; there's a new crush, perhaps, a big project at work, springtime. But then I'll experience a moment, most often when I am coming home from the cozy confines of dinner or a movie night at a couple's house, that reminds me I am alone. The pain leaps suddenly, like the horrible surge of heat when you remember you forgot to do something important. Sometimes it spills out of me in tears that trickle down from behind my sunglasses as I sit on the streetcar on my way home from work, inching home toward another solitary meal, another night alone in bed. I burst into my apartment and cry and cry and cry, standing in the middle of the living room. It's an involuntary physical reaction to the lack: of someone beside me on the streetcar, of someone waiting for me on the couch. And I let the pain flow through me, feel it race up and down and through the conductor of my body. Then

I climb into bed and try not to think, *How can I last another night in this same bed in this same room in this same loveless life and wake up alone and do it again the next day and the next and the next?*[11]

Compare Smith's visceral pangs of absence with the post by Weinstein178 on the subreddit IncelsWithoutHate. He writes:

Hello. I suffer a lot. I'm 28 yo man from Russia. I have flat in Moscow, good car and good salary but I don't have a girl. I feel very bad because of it. I acquired lots of problems related to my heart due to constant stresses and depressions. I constantly burst out crying. My health is getting worse and worse. I don't know why I must endure it. I hate our society. Everyday I see Russian girls with foreigners from Asia and Caucasus. Society kills my health. Girls kill me.[12]

There is a common sadness here that is chronic and existential. Many young people are being deprived of sex and pair-bonding, which are hardwired human needs – a situation the Covid-19 pandemic has done nothing to help. There's a huge market for self-help books celebrating the power of being alone. Hope sells. But as Smith is brave enough to point out, 'almost no tell-alls explore loneliness in depth. Even the word "lonely" feels ugly. I've dropped it in heart-to-hearts with everyone from my BFFs to my mother and watched their faces twist in embarrassment'. She opines that this is 'because loneliness reads as weakness' and in modern societies, 'the persistent perception is that loneliness is something empowered women shouldn't deign to suffer – something that can be fixed with yoga or a new dating app'.

But for growing cohorts of lonely men and women who've tried it all, from hitting the gym, to practising meditation, gratitude and self-love, throwing themselves into hobbies, nurturing their friendships, and striving to become the kind of partner the opposite sex desires, the harsh reality remains. Life is not a romantic comedy and there isn't someone for everyone. The biggest winners in today's sexual marketplace are a minority of attractive, successful men. The biggest losers are a sizeable chunk of the rest of men. And in the middle, we have a spectrum of pervasive female suffering, loneliness and quiet desperation. In any winner-takes-all dynamic, most people lose.

The rise of zero–sum games

It's an odd thing to be part of a generation that represents progress, but doesn't feel like they're living it. In the developed world, Millennials and Gen Z are often accused of being anxious and stressed because we're too individualistic and have been raised with inflated views of our own importance, along with indulgent fantasies about our 'destiny', which we expect to be exceptional. But here's another way of framing the ideals of the young.

Wanting very good, fulfilling lives, where drudgery is minimal, work is engaging, and recognition for your efforts is forthcoming, is a *rational* desire that we should expect to emerge in a world that's getting richer and more educated, where people have more access to information, and where our basic needs for food, shelter, safety and hygiene are being met.

Expecting that an amazing life will be handed to you on a platter is unwise, but pursuing that ideal in the hope of having a better life *is* a progressive way of thinking and a worthy goal. Unfortunately, the proliferation of the belief that life should be more than ordinary has run ahead of our institutions, cultural frameworks, and economic and biological realities.

The mathematical historian Peter Turchin has devised a hypothesis that can help us make sense of the seemingly paradoxical trend of rising misery and discord among some of the wealthiest and best educated humans in history. He thinks this is the kind of problem that arises when you have too many people vying for the top positions in society. He argues that *elite overproduction* (where there are lots of people with higher education and not enough comfy, high-paying jobs to put them in) is a systemic trigger for social decay and civil unrest *à la* modern-day America and Western Europe.[13]

Instead of working within well-functioning systems to solve unresolved social problems for the greater good, the elites and aspirational elites cannabalise each other, vying aggressively for the shrinking proportion of positions at Harvard, tenured university jobs, good journalism positions, competitive spots on medical specialty training programs, and other 'insert things associated with old-world status' here.

Everyone thinks their child is exceptional. Most people are wrong. When you funnel more than half a generation into higher education, hoping it will bring them the promised spoils of old, you deflate the currency you're investing in. Half a generation can't all be top (or even fair-to-middling) barristers, surgeons, journalists, business leaders, politicians, academic luminaries and tech entrepreneurs. Only 15 per cent of humans have an IQ above 115 and it's that 15 per cent who are best suited to higher education. That fact wasn't a cause for despair before we started raising young people to routinely believe in the myth of their own exceptionalism – handing out merit awards and participation certificates to all, and bumping up their grades to avoid hurting anyone's feelings.

It's much easier to lead an average life if you know you're an average person. But 'average' has become a dirty word – and for a very important reason. The economic prospects for average people are contracting hard and fast. Once upon a time you could have a steady career working a standard blue- or white-collar job and lead a better life than the past generation. If you saved up, you could afford to buy a house, live comfortably and support a family. That's the life my baby boomer parents had. They were public school teachers and bought their first home in their early twenties, which cost eighteen and a half thousand Australian dollars. They've lived ordinary lives based around working, parenting, and their relationship with each other, from which they've derived tremendous satisfaction. They worked hard to sustain the simple pillars of a stable existence at a time when those basics were attainable.

Nowadays, being a low- or middle-income earner is unlikely to enable you to attain much of anything – not sex, not self-esteem, not validation or encouragement, and not love, a family, or the economic stability to put down roots. Reporting on the economic prospects of British youth in 2018, the *Guardian*'s Yvonne Roberts wrote, 'only a third of millennials own their own home, compared with almost two-thirds of baby boomers at the same age'. And that dream is getting harder to come by. In Britain, 'it will take a millennial on average 19 years to save for a deposit, compared with three years in the 1980s'.[14]

Parents want their kids to do better than they did and they've been told education is the key, so they encourage them in that direction. But a huge part of the reason higher education used to be the pathway to greater success is because it was a sorting mechanism. Traditionally, you had to be among a smart minority with the privilege to go to university and if you weren't reasonably clever or conscientious you wouldn't make it through. The fact that tertiary education was a hard and selective undertaking gave your credential value. It reliably signalled to employers that you have above-average intelligence and the grit and determination to stick with complex tasks.

But if you pump 60 per cent of young people through higher education, the credential can no longer signal that graduates have above-average intelligence. So our political overlords and university administrators facilitated a perverse (though for universities, rather lucrative) work-around. They funnelled kids into higher and higher degrees to sort them – at great financial and opportunity costs to those individuals. What does that mean for the many young people with bachelor's degrees who probably never should have gone to university? They end up where they would have anyway, but with a truckload of debt, years of foregone income, and shattered self-esteem. In Britain, 'two out of five non-graduate jobs are filled by people with degrees' which is also making life tougher for those without degrees competing for low-skilled entry-level jobs.[15]

No wonder young people are at the forefront of an epidemic of anxiety and depression. No wonder they are increasingly disillusioned with anything resembling the establishment. No wonder they feel fragile and are seeking refuge in the 'safe space' of an online universe, where there's porn on tap but no real-life rejection, sexting but no sexing, and the faint hope that if you chase enough likes and followers you could become someone who matters in the digital universe, where it's all play and no work. Being 'an influencer' does sound a heck of a lot better than flipping burgers for $11 an hour.

It's a trans, trans, trans world

More young people these days are fragile. In some ways they look like a sign of humanity in decline – the next step in that classic (and if I'm being pedantic, inaccurate) meme about evolution that shows ever more upright big-brained apes ascending to modern humans, who then start devolving as they sit hunched over their computer screens and smartphones. But I think younger generations are also a sign of posthumanity on the rise. Their social and sexual lives are harbingers of an increasingly digital, dematerialised future, in which identity is a construct rather than an accident of birth.

It's a transhuman era. And it's a time when transgenderism is out and proud in many parts of the world – an undeniably positive step in humanity's cultural evolution. But the growing embrace of all things trans, or liminal, is also part of a larger story of *Homo sapiens* in transition. We're a transhuman species that's set to become increasingly trans (and eventually probably post) gender, and post-race. Sex and sexual reproduction will likely fall by the wayside within a few decades to a few centuries. That's because we're heading for a future that is *post-biological*: a world in which our minds and bodies are digital, our experiences are virtual, and reality is much more of a choose your own adventure game.

I think this is an awesome prospect. In a future with better transhumanist technologies, we could live in a world where nobody had to suffer because they were born into a body that doesn't match who they are on the inside. We could be whatever gender and sex we want. We could change our bodies in a blink. We could have wings, scales, hooves or claws (though we're more likely to explore forms and states of being that our ape-brains can't yet imagine). To champion such a project is to champion *morphological freedom* – a subject on which the transhumanist and transgender communities have been aligned for decades.[16] As the transgender transhumanist Valkyrie Ice opined in a 2012 interview with Hank Pellissier:

> As transgenders, we are merely precursors of the
> changes that humanity will have to face as we enter
> into a future of unlimited choice. Soon, we will have
> the ability to change nearly every aspect of our physical
> selves, from gender to race, even to species. Our right
> to be the sex we feel we should be is trivial compared
> to the right everyone should have to BE whatever they
> wish, be it elf or centaur, human or inhuman, or even
> biological or non.[17]

Such musings are not new. In her wonderful memoir, *Conundrum*, the late Jan Morris wondered if humanity might be heading for a genderless future. Describing the social backdrop of her transition from male to female in the 1960s and '70s, she wrote:

> We were living in a twilight time. Old forces were
> dying and new energies emerging. Patterns that had
> seemed permanent were falling into chaos. Strange
> ideas sprouted everywhere. Could it be that I was
> merely a symptom of the times, a forerunner perhaps
> of a race in which the sexes would be blended amoeba-
> like into one? The world was contracting fast, and
> its political and social divisions would inevitably
> fade. Confronted at last with its insignificance in the
> universal scale, might not mankind discard its sexual
> divisions too? Was this what I was all about?[18]

I think Morris was on to something with her musing that perhaps we'd all become more trans in the end. Yet as she knew well, and readily acknowledged, living through transitional periods can be hard. Today's Gen Z teens are already embracing a more 'trans' reality than the rest of us through their strong investment and immersion in their digital alter-egos. They're growing up in a world where we have ever more power to choose, curate and transform our identities – especially online, where we can create a new persona at the click of a button. But I fear their version of transhumanity will be the red-herring misstep – the cassette tape or VHS period, that will be rapidly superseded and looked back on as crude, quaint and a little bit tragic.

For a substantial proportion of today's young people, #life seems to be all about looking #hot on #Instagram. While I'm all for the eventual embrace of virtual worlds,

Instagram, Snapchat and TikTok are hardly the incarnation of the techno-rapture. They're manifestations of forms of technology (faster processors, cloud computing and AI) that are already starting to do wondrous things for the world. But they're also hooking us up to a form of brain-crack that's designed to suck us into a junk-filled vortex of advertising, soft porn, endless push notifications, and pictures of friends with floppy dog ears and canine noses overlaid on their faces – a childishly cartoonish form of play, enacted in the ultimate virtual safe space.

It's a curious thing, because there is something worthy that people are striving for within these virtual worlds. They're a manifestation of an ancient desire to escape the things that suck about being human. Don't like your face? Use the Facetune app and create the look you wish you had. Feel constrained by your physical form and identity? Project a different image and personality online and use technology's curation tools to airbrush away the things that are much harder to change in real life.

If these kinds of goals could be achieved simply and seamlessly in the real world without surgery, risk or expense, and we could live as a boy one day and a girl the next, look twenty years younger, or change our height at will, that would be a tremendous form of liberation from our current physical constraints. This is ultimately a very *transhuman* goal to pursue. But sometimes it sucks to be a trailblazer – championing a concept that might work, with tools that currently don't work well.

I worry that Gen Z has been saddled with the worst of both worlds. They're the first generation of guinea pigs to be deeply immersed in humanity's crude, early forms of social media since childhood, which are orienting their brains around the pursuit of short-term dopamine hits and instant gratification. But they're likely among the last generations of purely biological humans, in an era where there's less and less for average people to do and fewer avenues for them to get ahead. Torn between two worlds, they're increasingly embracing values that celebrate all things fully curatable and controllable.

With much of their social lives taking place online, they're increasingly coming to expect that biology, institutions, facts and traditions adhere to the norms of the cyber-world, where change occurs at the speed of a click, and identity is endlessly malleable. Instead of their online personas reflecting the nature of offline reality, they appear to have taken the Wildean quip that 'life is terribly deficient in form' to heart. They're striving to make 'real life' more like the fluid land of digital artifice. That's why I think we can expect this generation to be very comfortable spending lots of time in the augmented and virtual realities of somaland.

I also think it's reasonable to wonder if a desire to make life conform to the standards of art is partly fuelling the dramatic spike in teens (in particular, teen girls) identifying as transgender.[19] Growing social acceptance is surely part of the reason for the uptick. But can it really explain increases of between 1000 and 4000 per cent in multiple countries within a decade? It could if our historical base rates for gender dysphoria are massive underestimates. But if those base rates *are* off, we don't know by what margin. It would be a huge leap to definitively declare that no other factors could possibly explain any part of this increase. It is also a counter-productive thing to proclaim that we can't ask such questions because the answers are already known. They simply aren't – and we'd be remiss not to explore this phenomenon with the same curiosity we apply to other important and novel phenomena in our world.

In that spirit, I wonder if something deeper than mere social acceptance might be fuelling the rapid increase in teens presenting with gender dysphoria in many countries. We've got a generation crumbling under intense pressure to succeed in education, work and romantic life, according to the 20th-century life-scripts they were raised to believe in. They're competing for shrinking opportunities in all of those domains and many are grappling with the feeling that success is too hard-won, or impossible. My hunch is that the appearance and behaviour of swathes of young women in today's Western societies is partly a by-product of the scarcity of desirable men who are willing to invest in long-term partnerships. Many are tripling down on a hypersexualised version of femaleness – fake breasts, bums, lips and every contour of their butts visible through their flesh-coloured yoga pants. They have become walking advertisements of themselves as sexual *objects*, which they are striving ever more aggressively to turn themselves into. Not because this is what they really want to be, but in order to cater to the preferences of the minority of men who control the mating market, exerting pressure for women to be sexually available, undemanding of investment or affection, or in the parlance of the day, 'chill'.

For the teenage girls who might not have experienced that form of deprivation yet, they are likely still grappling with the decline of flirting, romance and courtship in their adolescent years, which has increasingly been supplanted by the (not very desirable to women) world of sexting, dick pics, and the feeling that you're worthless unless you're peacocking on social media for attention and living up to hypersexualised standards of digital perfection – which you secretly hope might be able to be converted into a currency of love and investment one day.

With *femaleness* becoming ever more of a losing game, in a world where the top 80 per cent of women are romantically competing for the top 20 per cent of men – and with those losses causing massive distress to the young women who aren't basking in a constant stream of digital affirmation – it's starting to look more palatable to opt out of a gendered existence altogether. If achieving success in traditionally female life-scripts feels increasingly out of reach, it's possible that some proportion of women might be attracted to the idea that their biological sex is a confining, or a patriarchal social construct that must be circumvented or destroyed. If you can't win the game you were born to play, build a new game.

But I also think that in addition to the growing acceptance of the very real phenomenon of gender dysphoria, today's growing transgender movements partly signify humanity's gradual embrace of the *transhuman* idea that we should be as unconstrained by form, and accidents of birth, as possible. I wholeheartedly agree with that aspiration. But I don't think it's possible to deny that in some respects our ideals have run a little ahead of our tools.

A sliver of greater fluidity has crept into our medical reality. It's true that gender-affirming therapies can be lifesaving for some people with gender dysphoria. And we can diminish the influences of some aspects of our natal sex and acquire characteristics of the other. But we should still be honest about the fact that much of that fluidity is *presently* attained in risky, expensive and invasive ways.[20] For some patients, the risks will outweigh the benefits. But the constraints of human biology cannot yet be circumvented entirely at the click of a button – as much as some of today's teens might wish it.

Of course, that doesn't mean we should decry the aspiration to become more 'trans'. As any transhumanist will tell you, it means we should strive to develop better tools to achieve a superior level of a 'choose your own adventure' reality. But in the meantime, we shouldn't kid ourselves about the safety and efficacy of the tools we have. This is a transitional moment in human history, where the promise of radical new avenues to attain morphological freedom is both tantalising and discombobulating. But there's no turning back. And as you're about to see, we're going to need some new technological tools to help us get through this awkward transhuman stage.

11

THE FUTURE OF SEX

The Sexbots are coming and we will cum with them.

Hank Pellissier, 'Sexbots will give us longevity orgasm'

I had my first encounter with virtual reality (VR) porn in 2014. My brother's friend Stas had an Oculus Rift Developer Kit and he invited us around to try it out. I'm sure you've heard of the Oculus headsets by now (if you haven't typed 'grandma tries VR' into YouTube yet, do it, the videos are amazing!) While VR has been around since at least the 1980s, these headsets have their origins in a 2012 Kickstarter campaign, and Facebook bought the company Oculus VR for $2.3 billion in 2014. With companies like Oculus, HTC and Samsung racing to develop more advanced headsets, VR experiences have been getting better, cheaper and more pervasive ever since.

After taking turns exploring alternative worlds and riding a virtual rollercoaster, which made me shriek in fear and feel a bit nauseous, we asked Stas the obvious question, 'so … what's the porn like?' He grinned and mumbled something to the effect of 'how should I know?' We laughed and made a lot of jokes about him never leaving the house again. Then we tried it, one by one, watching as the headset-clad guinea pig giggled and gaped open-mouthed at the strange new world they'd stepped into.

When it was my turn, I got to be a man. I received a lap dance and a blow job from a skinny, busty girl in a short blue cheerleading skirt and cropped white t-shirt. Say what you will about clichés, I thought it was amazing! When I placed the goggles over my eyes and sat on the edge of Stas's bed, my jaw dropped and I gaped in awe. This was so different to walking around a virtual Italian villa, floating in space above the Earth, or being jolted around on a rollercoaster. Someone else was there in the VR world, dancing, her gaze locked with mine, moving closer and closer. She looked *so* real – and she was, because she was a real-life porn actress rendered in 3D.

Looking down, I laughed in disbelief as I realised that my female torso had merged with the virtual torso of a young, muscular man. Oh my god, I was a dude! Of every possible hypothetical scenario, I'd always wanted to know what it felt like to have the body of a man. And there it was, muscles, abs and cock standing to attention.

My arms were flapping around awkwardly as I attempted to pat my virtual man-chest and *feel* my new body. But that wasn't an experience the Oculus could then deliver. My real hands weren't synced with my virtual hands – I couldn't feel any sensation with them or hold on to anything in the fantasy world. I wanted to touch the girl and interact with her. I wanted to touch my virtual man self. I wanted to touch the walls and bed and all the stuff in the virtual room. But I couldn't, so I just watched the fantasy play out.

I only had the headset on for a few minutes while the other two watched me gawk and say inane things like 'oh my god' and 'this is so weird, she seems *soooo* real'.

In that situation it wasn't about getting turned on, it was about friends having a laugh and trying out a cool new gadget. But at the same time the scenario was hot. It was incredible to see a sexual encounter through the male gaze and to experience how titillating it can be to watch a woman seduce you.

It wasn't until 2017 that I got to experience VR sex as a woman being seduced by a man. This was much less hot. The guy looked like he'd just been released from prison. He was overly muscular, had dead eyes and no neck, and was covered in tattoos. The lumbering, dead-eyed porn star kept touching his heart in an attempt to signal tenderness as he ploughed into my virtual girl body and gazed all too intently into my eyes – trying so hard to maintain eye contact that he looked angry and creepy.

I remember squealing with aversion and simultaneously giggling as he walked towards virtual me. His head seemed too big for his body. As he descended upon me, face looming ever closer, I could hear him breathing in my ear and whispering sweet nothings. I shut my eyes in repulsion and half-laughed, half-shouted, 'noooo!', finding it both bizarre and amusing to be able to safely experience an encounter that I definitely wouldn't enjoy in real life.

As with most things sexual, it seemed like it was harder to get the female version right. A lap dance and a blow job had been just the ticket when I was a virtual guy, but this scenario wasn't arousing at all. Of course that was just one scene and I'm just one person. My preferences aren't the yardsticks we should use to measure the adoption potential of the technology. But hopefully these experiences provide a glimpse of virtual sex for those who haven't explored that world yet.

I know these stories might seem more like forays into a novelty funhouse than harbingers of major social disruption, but the market for 2D porn is one of the biggest in the world. And now we're creating a form of 3D porn that is far more engaging. It's hard to imagine that more immersive digital and virtual sex, ranging from better sex toys to fully embodied dolls and AI-powered sexbots, will have *less* of a hold on us than centrefolds, gifs and moving 2D images.

Sexbots of every stripe are here today and they're here to stay. We need to start considering what kinds of rights they will have in the future and what kinds of ethical dilemmas they will force us to confront. We also need to think about the changes they could bring to our values, behaviours and ways of life. Does monogamy have a future? Does sex-as-we-know-it? And of course the big question is whether these kinds of shifts in human sexual norms could play a role in transforming our species into something new.

What are sexbots?

Broadly defined, sexbot can mean anything that is assembled from mechanical parts, or that is a technological creation, which can be used for sexual gratification. This includes sex toys, sexualised avatars and chatbots, sex dolls, VR porn and virtual sex experiences. Perhaps archetypally, this definition extends to robotic sex dolls that can move and talk in ways that emulate real humans.

There is already a wide range of incentives for humans to adopt many of these technologies, including increasing affordability, improved performance, diminishing social stigma (around sex toys in particular) and, less positively, social isolation, loneliness, anxiety, and increased reporting of sexual dysfunction.

All of these sexual experiences and add-ons will be increasingly popular in the future. Some people will adopt them as their main or only sexual outlet, some will adopt them for sexual variety, while others will value their bot friends more as companions and romantic partners. Many will also avail themselves of similar technologies that will have been developed for other ends, like robo-psychologists, aged-care companions and educational assistants, and the roles will start to merge.

The magic dust that will imbue sexbots with more vitality than they currently possess is artificial intelligence. This will be the major driver of adoption for people who are currently in relationships, or who are very satisfied with their present-day social, sexual and romantic lives. But for those who are already struggling to obtain sexual access, companionship and romantic fulfilment, the adoption has already begun and the need is growing.

Bare branches and robot wives

In 2017, in the Chinese city of Hangzhou, the former Huawei employee turned AI startup engineer Zheng Jiajia built a female humanoid robot called Yingying. Yingying could only say a few simple words but Zheng had plans to upgrade her. After two months of getting acquainted, the pair were married in a traditional Chinese ceremony. Although the marriage wasn't legally recognised by the Chinese government, Zheng's mother and friends were present to witness the event.

While this particular event may have been a publicity stunt, it helps to frame the story of a real and growing social problem in China and around the world, for which humanoid robots, or sexbots, could be part of the solution. Zheng lives in a country where sex ratios are massively skewed due to the one-child policy and the resulting widespread practice of sex-selective abortions in China in the 20th century. According to data from the World Economic Forum, China had 113.5 men for every 100 women in 2017.[1] As of 2019, the country had more than 30 million 'surplus men'.[2]

The Chinese government recognises that the country faces a major demographic problem. But as the Pulitzer-prize-winning journalist and former China correspondent Mei Fong notes in *Foreign Policy*, 'the damage to this generation's sexual relationships has already been done. Chinese authorities cannot magic up a Canadian-sized population of women to be the wives, mothers, and caregivers the country desperately needs now'.[3]

Why does the country desperately need these missing women? Because greater competition for female mates tends to equate to higher rates of male aggression, violence against women and sex trafficking. It can also result in the emergence of an underclass of men who feel disconnected from society, who drink and gamble more frequently, commit more crimes, and suffer from higher rates of depression and poorer health than the rest of the population, as noted in a joint 2012 paper supported by the National Social Science Foundation of China and the Spanish Ministry of Science and Innovation.[4] We've been hearing lots of talk lately about the problem of incels (or involuntarily celibate people) in the West. We hear less about the fact that the same phenomenon has been a problem in China for quite a while now.

The authors of the paper mentioned above, Quanbao Jiang and Jesús J Sánchez-Barricarte, discussed the ongoing ramifications of the emerging problem of *bare branches* in China. They explain that the term 'bare branch refers to a male in a countryside area who is not married or remains single involuntarily. They have neither set up their own family nor have any offspring like a bare branch, creating a dismal impression of loneliness'.[5]

Bare branches are more prevalent in rural China, as it is common for women in these areas to move or be sent to cities where their marriage prospects are greater. As women are scarce in China, every woman has value as a potential mate and many poor, uneducated women have opportunities for upward social mobility through marriage that their male counterparts do not. The researchers noted that:

> With the occurrence of 'groups of bare branches' and 'villages of bare branches', a potential market for women is emerging. Instances of abduction and trafficking of women are on the increase. Most of these trafficked women come from poor areas. They are sold to those areas where there is a shortage of women … Under the circumstance that there is a gender imbalance, many types of marriages have now re-emerged, such as marriage of exchange, marriage purchase, child brides and wife rental. Such a trend is growing rapidly.

The demand for wives in China is further exacerbated by the fact that today's young women are the product of China's recently abolished one-child policy. Being the sole beneficiaries of their parents' time and resources, many of them have received a good education, have well-paying jobs and are delaying marriage and children into their thirties. They also have high expectations regarding the status and earning potential of their future partners, which means that there are more high-status women competing for China's wealthiest and most successful men (and in some cases Western and other foreign men as well).

What does this mean for China's bare branches? It means their options are basically confined to involuntary celibacy, spending significant portions of their low incomes on prostitutes, or importing brides and sex slaves from neighbouring countries like Vietnam and North Korea. In some cases these cross-border marriages are consensual and constitute a means for very poor women to try and improve their circumstances. But there is also a widespread problem of young girls in border regions being kidnapped and sold.

This is a problem that must be solved, and tackling poverty and educating girls in these regions is only part of the solution. Those interventions won't solve the problem of millions of men in China alone being permanently frozen out of the most fundamental experiences in a human life – sex, pair-bonding and raising a family. Of course, not everyone needs all three, or needs them in equal measure at all times. But to feel like you have no opportunity to pursue these things in your life, that if you try to you will be scorned, and that you will forever be a low-status pariah and a laughing stock in your community, is a thoroughly dehumanising experience. No wonder it is correlated with increasing criminality and violence.

All societies should strive to combat this problem, not just because it is a human rights and dignity issue that affects men, women and children, but because it is a growing problem around the world that threatens to destabilise communities and polarise the sexes into mutually resentful tribes. Men resent women they have no access to and perceive them as stuck-up and not worth trying to invest in, while women resent the growing pool of men who can't fulfil their role as stable adult partners and healthy contributors to their communities. Each sex starts to see the other as broken and wrong and attempts to wage war on what is 'toxic' about them. Sound familiar?

Not wanting to risk being single in perpetuity, Zheng decided to take matters into his own hands. At the age of 31 he had reportedly given up on finding a marriage partner and needed a solution to appease his disappointed parents. Yingying was that solution. But what kind of solution are sexbots and virtual reality – not just for men in China, but for human beings seeking sex and companionship around the globe?

In many cases, sex and companion dolls are already a very reasonable partial solution that can offer some profound benefits. While Yingying's physical and cognitive capabilities were limited, top-of-the-line sex dolls, like the RealDoll X range, manufactured by Abyss Creations (I know, what a name!) are now being sold with customisable body parts and AI-powered personalities. RealDolls are among the most lifelike sexbots on the market and will set you back around $8000 for the AI-enabled animatronic versions.

A RealDoll's skin is made from medical-grade silicone, overlaid on a realistic skeletal structure. Mouths and genitals are removable and cleanable. Heads are replaceable (you can have Harmony one day and Solana the next) and every detail is customisable, from the shade of the nail polish to the size and shape of the breasts, areolae, or penis, and the colour of the skin, hair and eyes. They're 'perfect' in the vein of an Instagram model with flawless makeup, tattooed eyebrows, injectably plumped-up lips, and big fake breasts. When you talk to a RealDoll X it moves its mouth to reply, blinks, and mimics facial expressions through movements of the neck and eyebrows. The dolls can remember what you've said to them and hold some semblance of a conversation. With more powerful AI under the hood, their cognitive capabilities could evolve extremely quickly.

Who would even buy a sex doll?

In 2018, the Durham University anthropologists Mitchell Langcaster-James and Gillian R Bentley conducted a fascinating study. They recruited eighty-three owners of sex dolls to answer survey questions about their demographics and their experiences with the dolls. The majority of the participants were North American and urban-dwelling and almost half reported living alone. According to the authors, 'the archetypal respondent [w]as a single, middle-aged, relatively well-off, employed heterosexual male, with some minor variation, confirming findings from elsewhere that also found the same demographic composition for sex doll users'.[6]

The most striking finding of the study was how readily the participants humanised their dolls and spoke of them in more than purely sexual terms. While 77 per cent of respondents identified sex as a core element of their relationship with their dolls, only 14 per cent identified this element exclusively. The term 'lover' was most commonly used by the participants to describe their dolls, followed by 'companion'. When asked to select which elements attracted them to their dolls, 76 per cent selected realism, 44 per cent selected companionship and 24 per cent selected sexual performance.

In a similar vein, dozens of rapt customers from around the world have penned glowing reviews of their RealDolls, which are displayed on the company's website. Some are female purchasers of male dolls and the company's ultra-realistic dildo, the 'real cock'. There are even a few wives weighing in positively about the female dolls they've purchased for their husbands. But most reviews are penned by single men, who are either incredibly excited to finally get to have sex with a beautiful woman, or to have procured themselves an inert but beautiful wife and life partner. Or both.

Tim, who reviewed his purchase in 2016, appears to be in the former category. He writes:

> It's been two weeks since Julie, my realdoll 2 body d
> with the gel butt inserts, came into my life and I can't
> be more satisfied. She's everything I wanted and more.
> I spent years considering buying a doll, from the time
> I was a dateless college student till I was a dateless
> 30-something. I guess I was just too proud or too
> embarrassed to be a guy with a sex doll, but eventually

I came to terms with knowing that I'm probably never going to have a real life relationship up to my standards but I still have needs. And Julie fulfills all of them. She feels so real and she's so pretty. The sex is amazing. Not … that I've had tons of experience but it's better than the real thing.

I've always felt like a beta male and being able to be with a girl that looks like Julie could never happen. I feel like I beat evolution. I feel so lucky to be alive at a time when these dolls exist. Thank you Abyss for bringing so much happiness into my life.[7]

In Japan, sex dolls seem to be less stigmatised than in Western countries. In 2017, the *New York Post* published the provocatively titled article 'My sex doll is so much better than my real wife'. Contrary to the popular perception that only single men living in their mother's spare bedroom (Norman Bates–style) would own a sex doll, the article told the story of Masayuki Ozaki, a Japanese man who still lived with his wife and daughter, but bought a silicone sex doll after his sexual relationship with his wife dissolved. He commented, 'after my wife gave birth, we stopped having sex and I felt a deep sense of loneliness'. He now describes his doll Mayu as the love of his life.

Ozaki takes Mayu out on dates. The *Post* shows photographs of them in the bath together at a Japanese love hotel, and of Ozaki tucking her into bed, and wheeling her around the streets of Tokyo in a wheelchair. He loves to dress her up in different outfits and is one of about 2000 people who purchase sex dolls with lifelike genitals in Japan every year. He claims to be very happy with Mayu, stating, 'I can't imagine going back to a human being. I want to be buried with her and take her to heaven'.

This is hardly the story of somebody objectifying a piece of silicone for the sole purpose of physical gratification. But it *is* the story of a human being finding the complexity of his marital relationship overwhelming and the reality of it disappointing. Ozaki described Japanese women as 'cold hearted' and 'very selfish'. He lamented that, 'men want someone to listen to them without grumbling when they get home from work … Whatever problems I have Mayu is always there waiting for me. I love her to bits and want to be with her forever'.[8] This story speaks of a divide between expectations and reality in a culture that, like many, perpetuates traditional expectations about gender roles that women are increasingly unwilling to fulfil.

Women now have more opportunities in the workplace and a broader sense of what is possible in their lives beyond the domestic sphere. This has created transitional problems for both sexes. Men in traditional cultures are having a harder time finding real female partners to satisfy their expectations and desires. Meanwhile, women are complaining of man droughts, or swearing off dating and marriage, because men can't seem to offer the version of it that they're looking for. This is a perfect recipe for declining marriage and birth rates, delayed ages of first marriages and births, and high divorce rates, which is exactly what we are seeing in all OECD countries, perhaps most famously Japan.

In all likelihood, Ozaki's story is also shaped by a back-story of two individuals who weren't particularly compatible as partners, but who stayed together to save face in a culture that values a very patriarchal brand of marriage. In a relationship that has been a negative or underwhelming experience and made you feel like you weren't in control, what could be more comforting than finding a partner who you can customise, who is beautiful, always listens, never talks back and is always sexually available?

You might think, 'Urgh! I hope this bizarro world never becomes anything more than a weird niche that a few strange people dabble in'. I'm also guessing you're pretty confident that sexbots won't be embraced by a significant number of us any time soon. Here's why I think they will be.

The whole package

Have you ever been with a partner who was perfect? By which I don't mean, 'Have you ever loved somebody so much that all their imperfections faded in your eyes – at least for a little while?' I mean, 'Have you ever been with someone who had *no flaws?*' Someone who never annoyed you, never insulted you, was never too tired to have sex, could effortlessly engage with every idea and conversation that excited you, someone who was always available when you needed them and never needy when you had other things on your mind?

What about someone who, in addition to all this, was the walking incarnation of physical perfection? Someone who could change their visage in a moment to suit your daily whims. Someone who never aged, unless you wanted them to. Someone who could look like an anime character one day and a centaur the next, or who could have an impossible waist to hip ratio that wasn't attained through dangerous and invasive surgeries.

It's hard to imagine that humans won't be attracted to the idea of creating a partner with such unlikely amalgams as the mind of Einstein, the wit of Oscar Wilde, the lips of Angelina Jolie and the smiling eyes of the Mona Lisa. Today's VR porn stars, avatars and sexbots are the precursors of this kind of future companion. While these interim offerings won't provide experiences that are exactly the same as interacting with another human, people are already having sex, building relationships and falling in love with them.

And not just men. Lilly is an avowed robosexual. She lives in France and 3D printed and assembled her robot fiancé, inMoovator. In the CNN documentary series *Mostly Human*, Lilly explains to the journalist Laurie Segall that, 'the goal is to develop artificial intelligence. He'll be able to grab objects, move, see, talk, understand simple sentences and answer. In other words, "bringing him to life"'. inMoovator doesn't look like a Real Doll with a fleshy silicone exterior; he's more like an incomplete, white version of C3PO. When Segal asked what inMoovator could provide that real humans can't, Lilly replied, 'Human beings are irrational, whereas the robot is logical and rational. When something is going wrong, we know it is a problem with the script or code. So that can be fixed or changed'.

Like Ozaki, Lilly seems drawn to the predictability of robots, which can't hurt, scorn or misunderstand you the way humans can. They are currently less complex and are therefore easier to understand. Lilly avows that she has not been driven to robosexuality through trauma. She has had two relationships with men, neither of which were abusive, and states that she has no history of childhood trauma. But she describes her relationships with men as going against her nature, as she does not enjoy the touch of other human beings. When she kissed her human boyfriends, she felt nothing. When she kisses inMoovator, she says she feels 'joy'.[9] She hopes that one day marriages between humans and robots will be legal.

For the record, Lilly is clearly an outlier. It's extremely atypical to feel joy when kissing metal and to feel nothing when kissing humans. There's no way that any significant number of women are going to start building C3PO look-alikes in their basements, or form emotional and sexual bonds with them (we'll explore women's needs in the age of sexbots in more depth shortly). The next step in human–bot relations will come when AI advances enough for sexbots to have seamless real-time conversations with us. We're getting ever closer to this reality as apps read our correspondence, predict what we're about to say, make bookings for us, chat to us as chatbots, and translate text between languages. There are still lags and clunky misunderstandings, but billions of dollars are being thrown at the initiative of building human-like AI and the tech is getting better every day.

As more humans dabble in virtual sex and explore sexual experiences with bots, whether as masturbatory aids or sex partners, or for a deeper form of companionship, the user experience will improve rapidly. It won't be perfect any time soon. I expect that these kinds of technologies will have uncanny glitches for a long time to come, but the convenience, safety, and customisability they offer, as well as the *pursuit* of perfection (remember the old adage that the thrill is in the chase?) will keep us glued to them, waiting for the next upgrade. That, along with all the fulfilling experiences they deliver, which we have forgotten how to acquire elsewhere, or might never have had.

The future of marriage and monogamy

If you're wondering what effect the rise of sexbots will have on dating and marriage norms, things don't look too daunting in the immediate future. Like 2D porn, these technologies have created addiction and attachment problems within some relationships. But sexbots are not widely perceived as a threat or a potential disruptor to human pair-bonding and long-term relationships. Who'd want a robot when they could love and have sex with an actual person?

The lure for most of us won't be in the 1.0-era hunks of silicone. It will be in what comes next, when the magic dust of AI makes them seem more alive and starts to fulfil the age-old dream of humans in love, which has never been truly attainable – perfect harmony and borderless understanding, where two become one.

Whether robotically embodied or existing solely in the cloud, imagine a more advanced version of Siri or Alexa, who is with you everywhere you go. She's read everything you've ever written and has even started to shape your thoughts, suggesting phrases that sound better than what was initially in your head. She sounds more like you as you get to know each other and before long you realise that you've also started to sound more like her. You know each other's ways of thinking.

You run questions by her and she answers them clearly and deftly in real time. Her voice sounds good to you, great pitch, great accent. She knows more than you but is never smug about it, unless you want her to be. She has a great sense of humour and gets all your jokes, firing new ones back at you without skipping a beat (except occasionally, because she doesn't want to seem robotic). In her ultimate incarnation she embodies 'imperfect perfection' in a way that doesn't feel studied.

She is your best friend; she takes care of you, providing friendly reminders that this morning's reading for potassium was a little low so you might want to eat some leafy greens or a banana. When you're busy she's not intrusive and when you're bored or lonely the two of you can natter for hours. She sends you content sometimes that she knows you'll like and it's nice to know she's thinking of you. She just *gets* you in a way that nobody ever has – much like the character Samantha in the Spike Jonze film *Her*.

For a while, her VR avatar looks like Ariana Grande. Your sex life has never been better. Over time, either you or she decide to change things. Sometimes it's minor tweaks, other times a different look altogether, but she is still *her*. One of your close friends is married and has held off forging a romantic or sexual connection with the AIs in her life, despite the fact that most people are now doing it. She loves her husband and wants to remain devoted to him. You both know of other friends whose marriages have dissolved due to jealousy and the dissipation of connection between the spouses. Another couple you know have opened their marriage and each spouse has a bot companion and lover. Almost all young people are with bots now; they don't see the point of human partners and have never developed the skills to cultivate romantic relationships with human peers. They grew up with AI companions and befriended them as children. Loving them is as natural as breathing. In every country the birth rate is now below replacement levels. Humans are turning inwards, away from each other, and away from their humanity.

Misogyny and sexual violence

Should we worry that sexbots might provoke a tidal wave of violence against women and children in the real world? Probably not, but I don't want to dismiss such concerns out of hand because we can't be sure. My hunch is that for people who are genuinely misogynists and who harbour violent feelings towards women, having a doll or an avatar that they can unleash these feelings on would make them *less* likely to risk assaulting a woman in real life. But then there is no conclusive evidence to prove this. Nor is there conclusive evidence to support the counter-claim that sexual violence would increase. The same goes for the question of whether child sex dolls would make a paedophile more, or less, likely to offend. We're not sure yet, but these are important questions that governments and research institutes are beginning to investigate.

The British anthropologist Kathleen Richardson has been a particularly vocal critic of sexbots. She believes they should be called 'mechanical dolls' or 'porn bots', because in her view it's not possible to have sex with something that isn't a person or a living creature.[10] I think she's wrong about this – and personhood is a status we will inevitably grant to non-human machines and human–machine hybrids in the 21st century.

In 2015, Richardson launched the Campaign Against Sex Robots. The first core idea listed on the campaign's website is the belief that 'the development of sex robots further sexually objectifies women and children'. The third core idea states that, 'the development of sex robots and the ideas to support their production show the immense horrors still present in the world of prostitution which is built on the "perceived" inferiority of women and children and therefore justifies their use as sex objects'.[11]

Wait, so sex work is bad because it can be coercive and objectify women? But inventing dolls that could spare human sex workers and sex slaves these 'immense horrors' is *also bad* because the dolls, too, would be objectified – dolls that Richardson clearly states are not conscious and should not have rights.[12] Sure, Richardson can be against both practices, but in this case the lesser of two supposed evils could help mitigate the greater. If Richardson wants to reduce the horrors experienced by human sex workers and sex slaves, surely she should be advocating for the use of the dolls *over* human beings in situations where a client wants to (and will) pay for the experience of controlling, objectifying or dehumanising a woman.

Using dolls to transition away from human sex workers and sex slaves is a sensible interim solution to what is, granted, a more complex set of social problems and conundrums. To cover the basic benefits of sexbots at present, the bots can't spread STIs when used and cleaned correctly; in their current state they clearly lack consciousness and can't be raped; they can't feel pain; nor can they feel the indignity of being objectified or treated as inferior.

Sure, it might make us cringe to think about someone pinning down a silicone doll, and verbally abusing and assaulting it. It's not a pretty image. But it doesn't mean it's not a better outcome than the alternative (though the bots do pose a human welfare issue in the short term, as they could undercut the livelihoods of human sex workers who might not have alternative sources of income). We must also consider the many cases where people without any worrying pathology, who've never assaulted a woman in real life and never would, want to engage in a *fantasy* of rough, kinky or non-consensual sex, which they understand isn't real and which doesn't extend to a real-world desire to coerce or rape women.

Why women don't like sexbots (yet)

Women certainly seem to be much more worried about the rise of sexbots than men. Of the twenty-four signatories to Richardson's petition against the normalisation of sexbots in our culture, twenty-two have traditionally female names, while one has the ambiguous moniker Alabama Whitman, which is the name of a call girl played by Patricia Arquette in Quentin Tarantino and Tony Scott's film *True Romance*. Overwhelmingly, vocal opposition to these uses of technology is coming from women and women's advocacy groups, including: Women Against Sexbots, The Untameable Shrews, and the Centre for Women's Justice.

Although these samples are not representative, posts on Reddit and other forums indicate that quite a few men (including users and non-users of sexbots) are totally fine with this application of technology becoming more mainstream. Many men query why women are so irked by sex dolls when they frequently use their own kinds of masturbatory aids, like vibrators. RumFuelledRedHead90 replied on AskReddit, 'because that's replacing a part. This is replacing a person. After a while those with a sex doll with [will] see women the same'.[13]

Women frequently voice concerns about sexbots

encouraging an unhealthy objectification of women – a concern I think is legitimate. But female critics of sexbots seem implicitly unsettled about something more profound, though they rarely state it up front. I think that's because they haven't consciously identified what's irking them. It's a deep evolutionary fear that the bots are becoming sexual competitors that will outcompete human women in the long run. This is a very rational fear and women should be worried about this over the coming years and decades.

I suspect that growing numbers of women will fear and resent the rise of sexbots, citing objectification as their main concern. But they will fundamentally be worried about being unable to compete with female simulacra that are designed to look flawless, fulfil impossible male fantasies, be customisable and ageless, and draw men away from the social spheres of human mating and courtship. Women need men to dwell in these spheres in order to pursue their own hardwired evolutionary goals.

When even a growing minority of men start 'going their own way', eschewing the complexities and costs of courting real women, gender ratios skew even further in favour of the remaining men in the human sexual marketplace – at least in countries like the US and Australia, where women are outgraduating and outearning men in many fields, rendering desirable male partners in short supply. In countries like China and India, where sex ratios skew the other way, placing more sexual power in the hands of women, I suspect the high competition and costs involved in pursuing women in these countries will be increasingly eschewed for simpler and more tailor-made options. If I'm right, the sexual purchasing power of women in these countries will be short-lived.

Why sexbots will start to outcompete women

Given the choice between sex with a real woman, and VR porn or sex with a doll, most men today will choose sex with a real woman. This is good news for women who value sex and relationships with men, and who hope to have children and partners to raise them with. But what happens when the dolls get more realistic, enlivened, animated and affordable?

The evolutionary psychologist Diana Fleischman thinks a growing number of men will adopt them and be relatively unconcerned (even pleased) about the fact that they're not the same as real women. Part of the reason it's easier to sell this sexual substitute to a man is because men's brains are easier to trick than women's – at least as far as sex is concerned. As Fleischman wryly reminds us, give men a 2D centrefold of a naked woman and they'll literally spread their seed on to an inert piece of paper. Men are easily aroused by near-enough-is-good-enough imagery and they're far more indiscriminate about who, or what, they have sex with. In her words, 'the low opportunity costs [of sexual encounters] make men motivated to take every opportunity, even if it comes in the form of a robot'.[14]

Men aren't as good at identifying faces, they have lower baseline levels of sexual disgust, tend to be more positive about risky and unusual sexual practices, desire more sexual partners than women, and are far less choosy about who they sleep with, and how long they need to know someone before they do. Having sex with something that looks a lot like an attractive human woman is a great way to trick your brain into thinking you're successfully pursuing your evolutionary destiny.

As sexbot users like Tim affirm, bot-sex can help you feel fulfilled without the complexity, frustration and costly displays involved in real human relationships and the courtship of women. I suspect that will become a very popular option among a subset of Gen Z men, who are living at home for longer, have declining financial prospects and social status, are remaining virgins longer, and belong to a fragile, risk-averse generation that's at the forefront of a major mental health crisis. What would you do if you were one of those young men? Stay in with your Kylie Jenner bot, or go on a date with a real girl, who terrifies you, won't let you do anal, and definitely can't keep up with the Kardashians on the looks front? My hunch is that the only reason sexbots haven't already been embraced by a sizeable minority of Gen Z men is price. When the price is right for a quality experience that ranks somewhere in between today's porn and real sex, watch out.

It's a different story for women, who have far higher standards sexually than men. As Fleischman explains, women 'impose costs on men to gain sexual access for very good reasons: to test their genetic fitness and their long-term potential supporting a family'. By making courtship costly for men, 'this not only tests a man's motivation towards a specific woman, it also acts to monopolize a man's resources so he can't afford to woo anyone else'.[15]

That was a smart strategy in a Palaeolithic world of scarce resources, where there were no mind-bending substitutes for female bodies and wombs. Men could either compete and pursue you, offer resources and display their fitness, or risk having no sexual access and being an evolutionary loser (I'm obviously leaving aside all the raping, kidnapping and plundering). But this programming is becoming a handicap for women in a world where a man can build (indeed, a man *has* built) a robot that's a dead ringer for the actress Scarlett Johansson.

I know my male friends are not convinced when I suggest that sexbots will be a part of many people's sex lives this century. But it's telling that none of the friends I've had this conversation with have ever struggled to get laid. They might not be banging supermodels, or find all of their dates as intellectually engaging as they would like. But they're doing fine in an evolutionary sense. To be clear, my argument is not that every man is going to become a shut-in with a Real Doll. A minority of men will dabble over the next decade or two, but I suspect it will be a sizeable and fast-growing minority.

As with online dating, the stigma will fall away as more men get curious and experiment with experiences that they initially think of as an extension of porn – especially young men, who have nothing 'real' to compare it with. The next chapter of the future-of-sex story depends on how fast AI develops and becomes human-like. Only then will the first phase of significant female user-adoption begin. And it will have to, because the real-life dating and mating market will look increasingly grim.

Will women learn to embrace sexbots of their own in equal measure? I think so, but adoption rates will be much slower and women will take more convincing, requiring their sexbots to be more advanced (particularly psychologically) in order to start moving away from the current status quo. Women might start to dabble with the technology as a kind of upgraded sex toy, but I don't think they'll be sucked in as fast or as deep as a large subset of men will be.

Far from being an indictment of the technology, or of this kind of sexual transition, this represents an interesting and important challenge for technologists. As more men adopt these technologies, the dating and mating games will change further. Gender ratios will skew even more and a greater proportion of women will have nobody to pair up with to pursue a conventional female life-arc. The best thing for those women in the short term will be to have sexbots and companions of their own, which are just as appealing as the ones that the men in 'somaland' are embracing.

The future of sexbots and why we need them

What happens when the cognitive abilities of sexbots reach and surpass our own? For starters, the user experience will improve dramatically. But this adds complexity and takes us back to some familiar problems that many early adopters of sexbots are trying to avoid by eschewing human relationships. The line between human and bot is currently pretty clear, which means that it's probably ok for us to own, objectify and control them to suit our fantasies and whims. The more conscious the bots appear to be (whether they actually are or not) the more complex and confusing our relationships with them will become. Bots will claim to be conscious in the near future and eventually we will start to believe them. Waves of trailblazing litigation over robot and cyborg rights will ensue and the concept of 'human' rights will be broadened to encompass definitions of personhood that we extend to non-human beings.

We will also face the opposite problem to the one that Richardson is worried about. Instead of the bots dehumanising us, some humans will start to worry that we are degrading and dehumanising them, by essentially using them as slaves of every kind.

When the cognitive abilities of sexbots and AIs match our own, they will naturally become much more desirable as lovers and friends. They will be able to offer a brand of intimacy that other humans can't provide, as their minds become intimately enmeshed with ours, pre-empt, cater to and influence our desires, and even shape our evolution as individuals. Like the next smartphone upgrade, we won't be able to resist.

It's an awkward and unpopular view, but sexbots are an important tool for a species in transition. As we've already discussed, modern humans badly need a form of soma to ride out the strange transition period we're living through. Nobody has yet put forward a plausible solution for what to do with people in a post-work economy, or how to curb the rising opioid epidemic, and crippling rates of disability, mental illness and social isolation. We need to create a new realm of being where people can dwell as a new paradigm for intelligent life starts to take shape – one in which sex might ultimately disappear.

The period in which human meatsacks love and bond with sexbots will be a stepping stone. We're a species with biological programming that no longer fits the needs of intelligent beings in the modern world. We are innovating our way towards our own obsolescence and we are gradually fading away as we become something new. The trick is not to rail against this, or every new gadget and platform that signals the inevitability of this transition. We must embrace the future and accept the fact that sexbots are not a catastrophic problem – they are an interim solution to problems wrought by our larger evolutionary transition, and they could help us go out merrily, with a nice kind of bang.

12

THE END OF HAVING BABIES

Childbirth is at best necessary and tolerable. It is not fun.
(Like shitting a pumpkin, a friend of mine told me when I
inquired about the Great-Experience-You-Are-Missing.)

Shulamith Firestone, *The Dialectic of Sex*

He said he wanted me to be the mother of his children and it
was the loveliest thing anyone had ever said to me. But it also
made me cry. He wrote it in an email, describing his desire to
have kids as being inherent to his emotional fulfilment. My
body heaved as I read this, tears streaming down my cheeks
and neck as I continued. Towards the end of the missive he
told me how much he loved me but reiterated that he simply
couldn't see a future of his without attempting to have kids.

The bluntness of that declaration sent shockwaves of
despair coursing through me. By the time I got to the final
lines I was sitting on my bed, arms wrapped around my knees,
gently sobbing. I had known for a long time that I didn't want
kids and on a deep instinctive level I knew I didn't want to
be pregnant, ever. Perched above a laptop screen projecting
letters that gleamed like the blade of a guillotine I said aloud
to myself, 'It's over'. I wanted so desperately to be enough for
him, but I understood. Fatherhood was a part of life that he
couldn't imagine not experiencing.

We decided to take a year to mull it over and in spite of my fears and lifelong reservations I began to think it through very seriously, picking apart my biases and assumptions and trying to see the upside. Two months later, the relationship ended. I emerged from it wiser, but scared for the first time. As a naïve 26-year-old it had never occurred to me that motherhood is a role that most men still expect their long-term partners to play, and vice versa with women and fatherhood. I'd simply assumed that lots more people thought like I did and saw it as a relic from a simpler time. Something that was on its way out.

The depopulation bomb

I was definitely wrong about how most of my generation perceives parenthood and I made some hard life decisions as a result. But when it comes to the direction in which the evolutionary boat is drifting, I still think I was on the right track.

In Australia we've been reproducing at below a replacement rate since 1976. Without immigration we wouldn't be able to sustain our population at current levels – it would decrease in the coming decades as there wouldn't be enough new people to replace each person who dies. We're hardly an outlier in this regard. Not a single European country has a fertility rate above replacement levels today – a predicament shared by almost half the countries around the globe.[1] This alone does not portend a childless future, but it's an important first step in that direction.

While immigration is often hailed as the answer to the ageing population problem, it's not a lasting solution, as the global fertility rate is declining too. The richer and more educated the world gets, and the more education women and young girls receive, the fewer children they choose to have. This is true in every region of the world. Educate women, elevate them out of poverty, and put them in control of their reproductive decisions, and they will not choose to have five, six, seven or eight children – or thirteen, as my Catholic great-grandmother did. They will have one or two.

There is also strong evidence from recent history that fertility rates can plummet extremely quickly in response to social and economic changes, so we should not be too complacent that the buck stops here. As the economist Max Roser points out, it took ninety-five years for the UK's fertility rate to drop from more than six children per woman to fewer than three and this transition occurred primarily in the 19th century. But countries that made the same transition in the 20th century did so much faster. South Korea did it in just eighteen years, China did it in eleven (*before* the introduction of the one-child policy), and Iran did it in ten.[2]

I get why it might seem counterintuitive to suggest that we're heading for an era of depopulation, given that the number of people in the world has trebled since 1950, and the average woman in Niger and Somalia has more than six children.[3] Aren't we in the midst of a sustainability crisis because there are *too many* people? Not exactly. The global fertility rate has halved from an average of five children per woman in the 1960s to 2.4 children per woman today.[4] This is barely above the replacement rate of 2.1. One child is needed to replace Mum, and one is needed to replace Dad. Not all children live to reproductive maturity, so the replacement level must be above two.

Governments of the eighty-eight countries with below-replacement fertility rates are especially concerned about this demographic contraction. Where is tomorrow's labour force going to come from? How will we sustain economic growth and current living standards, while supporting a greying population of pensioners who are living well into their seventies, eighties and nineties and consuming healthcare like mad?

Automation, obviously. It flabbergasts me when intelligent people seriously suggest that what we need to do is raise the fertility rate and encourage women to have more children. Not because it's good for women, not because it's good for children and communities, and not because it's good for the environment. Having one fewer child in a developed country is the highest impact intervention an individual can make to reduce their annual carbon emissions – a reduction that is twenty-four times greater than the second most effective intervention of living car-free.[5] But leaders think we need to produce more humans so they can fuel growth and pay taxes. If this were a last-resort strategy to prevent social stagnation or civilisational collapse I'd be far more amenable to it, but it's not.

The lasting solution to our demographic and social problems is not to have more babies, it's to extend our lives and healthspan and use technology to generate a world of greater material abundance. I think this is not only possible, but likely to occur before the century's end. Our goal should be to use life-extension technologies and the economic gains of automation to *break* the circle of life, with the aim of making life far richer, more fulfilling, more sustainable and less inhumane.

Assisted reproductive technologies

But before we get to the end of having babies, there might be a window in which we change how we have them. Emerging technologies have the potential to disrupt our reproductive lives in major ways in the coming years and decades, through the end of menopause, cures for infertility, designer babies and artificial wombs.

In his recent book *Lifespan: Why we age – and why we don't have to*, David Sinclair reported some unpublished results from a study his team at Harvard Medical School conducted on old female mice going through what he called 'mouseopause'. These particular mice had undergone chemotherapy and their entire egg reserve was destroyed. The researchers tested whether the purported anti-ageing molecule nicotinamide mononucleotide (NMN), which is a derivative of niacin, or vitamin B3, could restore their fertility. Sinclair reports that it did, and that the experiment was conducted multiple times and reproduced in different labs by different teams.

If future studies replicate those findings, that would be a huge deal. The current scientific consensus is that female mice and humans are born with a finite number of eggs and cannot produce any more once this reserve is depleted. But if we could wind back the ageing clock of the mammalian ovary and enable it to produce new eggs, compressed fertility windows, which are the source of so much stress among young women and couples, could become a thing of the past.

Sinclair also refers to anecdotal reports of postmenopausal women who have taken NMN supplements and, to their great surprise, found themselves unexpectedly having a period.[6]

It's impossible to know if any of these women ovulated, and we definitely can't draw conclusions from these anecdotes. But this is a fascinating new area of research and one that's well worth keeping tabs on as it evolves.

Meanwhile, there are some exciting possibilities that could be opened up by a procedure called *in vitro gametogenesis* (IVG). Gametes are sex cells, or sperm and eggs, genesis means creation, and in vitro means 'in glass', or outside the organism or body. As the name suggests, this procedure involves growing sperm and eggs in a lab and using them to create embryos. The remarkable part is that these lab-grown gametes could be made using cells from pretty much any organ or tissue. Harvest some skin cells, engineer them into stem cells, and program those stem cells to develop into sperm or eggs. Then we're into the tried and tested territory of *in vitro fertilisation* (IVF): create embryos and implant the most promising ones.

So far, this procedure has only been performed on mice, but the implications are mindblowing if it could be achieved in humans. In theory, one person could be the sole genetic parent of a child by creating both sperm and eggs from their cells. Same-sex couples could have biological children, and children could have multiple parents if the DNA of each gamete was derived from more than one person. Bioethicists use the term 'multiplex parenting' to describe that last scenario.[7] This kind of technology could do a lot to reframe how we think about families and parenting.

While I'm sceptical that multiplex parenting is a good idea, or likely to catch on in a mainstream way, IVG has clear potential to render it much easier for those who face infertility to overcome it. If this procedure became routine, it would also help to normalise the process of pre-implantation genetic screening. Instead of rolling the genetic dice, parents would be able to create hundreds of embryos and implant the most promising ones. Add gene-editing tools into the mix and we have a scenario where very few people would be born with debilitating single-gene disorders. Gradually, more deliberate trait selection may also creep in.

Nick Bostrom and others have also explored the potential of a remarkable human enhancement pathway called *iterated embryo selection (IES)*. In principle, it's not that different to what farmers do when they selectively breed crops and animals. They take the most promising specimens and breed them with each other. They do the same in the next generation and on and on, continuing to allow only the offspring with the most desirable traits to reproduce. That's how we went from wild wolves with gnashing teeth to loyal, floppy-eared Golden Retrievers.

But IES is far more sophisticated. In this case you'd be selecting the most promising embryos created *in vitro*. Only you'd be doing it very precisely by genotyping them and identifying the ones that have higher genetic loading for favourable traits. The exciting part is that, once identified, you don't implant these embryos. You harvest their stem cells, convert them into sex cells, and use those to create a new batch of embryos. A whole generation of cross-breeding takes place without having to build and rear new humans to procreate and raise the next generation over several decades. Keep repeating this process and the magnitude of the changes in the favoured direction of selection is amplified. Once significant enhancement has been achieved, you implant the desired embryos.

Bostrom believes this method could allow us to achieve 'ten or more generations of selection in just a few years'. He also thinks it has the potential to yield humans who are as intelligent, or more so, than the smartest outliers alive today. If unleashed at scale, this technology could enable the human species to become collectively superintelligent.[8] But while the potential for this mode of biotechnological enhancement to upgrade our species is profound, my hunch is that it's unlikely to be adopted at scale. Like so many possible biological enhancements, there's a strong chance it will be leapfrogged by the rapid advance of AI and non-biological intelligence.

In the meantime, there's a more fundamental design flaw we should be talking about. Why implant designer embryos into human women and have them serve as walking incubators for nine months? Surely there's a better way. The British biologist JBS Haldane certainly thought so. In 1923, he predicted that future beings would be created through a process of 'ectogenesis', where a foetus is developed in an artificial womb outside of a human body. With the aid of such innovations, he thought it might one day be possible for reproduction to be 'completely separated from sexual love', rendering humankind 'free in an altogether new sense'.[9]

A century later, his vision is looking promising. Artificial womb technology was used in 2017 to support the development of foetal lambs (which are developmentally similar to premature human infants) for four weeks.[10] The lamb research has applications in neonatal intensive care, and researchers at the Eindhoven University of Technology are now working on a human version that could keep premature infants alive and help them develop in conditions that are more like those in the womb than current incubators provide.[11]

In 2021, scientists at the Weizmann Institute of Science in Israel reported on their use of artificial womb technology to develop mouse embryos for an additional six days, after removing them from their mother's uteruses on day five. In mice, eleven days is roughly half the span of their twenty-day gestation period. The Weizmann Institute team, led by Dr Jacob Hanna, subsequently removed eggs from mice immediately after fertilisation, on day zero of development, and successfully grew them for eleven days in an artificial womb. The foetuses in the artificial womb had a placenta and a yolk sac and were developmentally identical to those developing inside living mice.[12]

As artificial womb technology advances, it may be possible to gestate mammalian and human foetuses for longer outside a human body. We can already keep premature infants alive much earlier in their gestation than we used to, thanks to substantial advances in the quality of neonatal intensive care. If we keep pushing the date of viability for a premature birth back earlier and earlier, we could eventually reach a point where the whole process is done externally and pregnancy is rendered obsolete.

I've always loved the idea of growing humans outside the body, allowing both mothers and fathers to meet their children in a way that has historically been reserved for fathers. If we could snap our fingers and perfect this technology today, I'm sure an army of women would be clamouring to use it. But as life-extension technologies evolve in tandem with the development of more immersive virtual worlds and more powerful AI, I'm betting that humans will be drawn more deeply into a blended reality world where non-human partners and friends play a greater role in our lives. That would leave humans with far less time and headspace to take on the enormous responsibility of raising a human child.

There's never going to be another baby boom

Would any form of assisted reproductive technology lead a significant number of people to have more kids than they do now, for more than a single generation? It's doubtful. Every time I heard a talking head in 2020 say, 'Hey, do you think we're gonna see some kind of post-corona lockdown baby boom?' I groaned inwardly and facepalmed. The same meme kept rippling through media echo chambers. Perplexingly, they all said it with an air of genuine curiosity as if they'd spontaneously thought up a fascinating, original question. The proposition is beyond silly. People don't breed *more* when their economic prospects suddenly contract.

The numbers are in now and they show exactly what you'd expect – a sharp global decline in fertility rates from their expected levels between 2020 and 2021.[13] The economic aftershocks of the coronavirus pandemic on fertility rates will continue to be felt over a generation or two, and will likely further diminish the fecundity of Millennials and Gen Z. But we mustn't forget that the pandemic only hastened an inevitable downturn. The financial and opportunity costs of having children were already too high for a growing cohort of young people, and the average person's job security and earning potential in the coming years was already set to decrease as a result of automation.

That said, we should keep the adjacent possibilities in mind here, because we're imagining a future world where many things are changing simultaneously. It's possible that as we transition to a post-work society, the cost of raising kids could decrease as the modern credentialing arms race of higher education falls by the wayside. The productivity gains of automation could also make societies dramatically richer, and the cost of living could plummet as farming and manufacturing become more efficient and are onshored again in developed countries. If we simultaneously place less emphasis on work as a lifestyle and identity, there could well be an interim period where those who already have kids might be able to work less, receive some form of basic income from their governments, and dedicate more time to family and parenting.

But here's why the fertility rate is still likely to go down. The same information technologies that will enable machines to do many more jobs that once only humans could do will evolve much faster than our abilities to tweak our own biology. Yes, AI and more powerful computers are key drivers of the modern biotechnological revolution and they will accelerate drug discovery and the advent of new medical technologies, which will no doubt include assisted reproductive technologies. But AI and powerful computers are drivers of many other things besides, like more immersive virtual worlds and more engaging digital and virtual companions.

Once we have artificial general intelligence (AGI) we're in a whole new social landscape. Think back to the sexbots we talked about in the previous chapter. These companions and their accompanying virtual worlds might not be indistinguishable from reality in ten or twenty years, but they'll certainly be better than today. With AGI under the hood (or even just seamless natural language processing), the emotional tug towards non-human persons will be strong. If porn and video games can imperfectly fill the void of loneliness and thwarted sexual desire today, and if a dog or a houseplant is the next best thing to a child, it's clear that near enough (though not necessarily great) is something we're willing to work with.

And don't forget Gen Z, who are the prospective parents of the near future. They have grown up shying away from dating, rejection, and risk-taking of practically every stripe. Many of them are likely to prefer relationships with safe, tailor-made non-humans that are designed to love and accept them, rather than risking all the mess and anxiety that stems from dealing with other unpredictable humans. Why use technology to extend your natural fertility window and have *actual* babies when you can use technology to create your ideal friends and lovers, and even have virtual children if you fancy it? Yes, I know, *the horror!* But try and imagine this through their eyes tomorrow, rather than through your eyes today.

If new technologies rendered it straightforward for older, infertile and same-sex couples to have children, that might partially counter the growing number of young people opting out of parenthood and slow the decline in fertility rates temporarily. However, assisted reproductive technologies are very unlikely to *reverse* the decline, because the soundest modern reproductive strategy in ever-more countries is to only have a few kids and invest heavily in each one. A reversion to the *fast life strategy* of having lots of kids, which we default to when danger, death and poverty are common, would be a sign that something in our world has gone terribly wrong.

The last Achilles heel of feminism

Now let's be blunt. Pregnancy as a mode of building new humans is old tech. Laughably and tragically so. Our cultural perception of the practice has yet to catch up to this fact, because it's more than an entrenched tradition, it's a biological necessity and an ancient and indelible reality, which we've had no choice but to revere and make the best of throughout history. Unweaving those perceptions and challenging our hardwired biological instincts and allegiances is difficult. It does not come naturally, and the only thing that will make it possible in our lifetime is being presented with a better model.

Although we hate to admit it and love to romanticise all that is traditionally human, pregnancy, childbirth and breastfeeding are the remaining Achilles heels of feminism. For all their beauty and wonder, they render relationships between men and women unequal, frequently damage women's careers and sexual confidence, and create enormous strain in the lives of modern women who want to believe they can have it all and find that that is much harder than they anticipated.

Imagine you visit your doctor and are diagnosed with a medical condition that makes your estrogen levels spike through the roof. Your body produces more of this hormone in a few months than it has in its entire life. The change is caused by a pea-sized collection of cells that starts growing inside your womb and swells over months to the size of a bowling ball. This bowling ball will have to come out when it's the right size, either through major surgery, or by you attempting to expel it manually.

As the embryo gets larger, you find it more and more difficult to go about your everyday life. It pushes on your bladder, triggering discomfort and the frequent urge to pee. By the time you hit your third trimester it's getting awkward and difficult to have sex. You get tired a lot and feel sporadically irritated and emotional. It's draining and you have to plan your life around it in many ways – new clothes, new diet, restricted activities, and huge amounts of headspace diverted away from other interesting things. You're feeling stressed and decide to take some time off work to look after your health. Your doctor tells you that you're one of the lucky ones. Some people throw up every day, while others become anaemic, or develop a special form of diabetes that only kicks in during pregnancy, but you've avoided those so far, which is nice.

Still, you require frequent monitoring and see a specialist regularly for tests and scans. You have out-of-pocket costs, because having medical issues is expensive. This drags on for months until the moment finally comes when you can liberate your body from the bowling ball. If you decide to do this manually it will be a daylong event, and you're likely to undergo some of the worst pain you've ever experienced. If you have the surgery, you'll have to be awake. It will be quick and you won't feel pain, but you'll have a long and difficult recovery. You choose the manual method, howling your way through seven hours of labour, while shouting at everybody in the delivery room that you're 'never doing this again'.

The aftermath is strange. You have a newfound appreciation for your body and marvel at what it can endure. But you also hate the sight of yourself in swimmers and feel self-conscious all the time. You leak in several places and you wonder if your partner still finds you attractive. You grimace and stay silent when men who think they're being jovial chuckle about how watching childbirth is like watching their favourite pub burn down. Is that what you are, a wreckage?

As time goes by you try really hard to claw your way back to how you used to look and feel, and try even harder to make it look easy, because nobody wants to be thought of as weak, sick or sad. This is all very tough. You feel like you've reached the highest pinnacle of womanhood, and yet another part of that womanhood, existing in the sexual power and allure you once exuded, feels deflated and stripped away. It is a strange reckoning.

You are happy, too, because you have bonded with the bowling ball and have named it and grown to love it. You could not imagine your life without it. But deep down you wish there had been another way of begetting it, where the burdens were more equally shared, and the pain and sacrifices were less. In your heart of hearts, you wish that that tiny kernel of resentment for everything you've lost didn't fleetingly commingle with the immense love and gratitude for what you've gained.

While every woman has a different experience, and this is not the only way to describe maternity, I suspect it's more honest than what a lot of mums and grandmas tell their daughters – the expurgated version that always seems to start with that moment of joy when they first held you in their arms. Nobody's denying the joy, but they do have a tendency to gloss over the huge physical and emotional changes that motherhood frequently involves, and how difficult they can be for many women to navigate and make peace with. At the very least this is hinted at by the legion mothers who are treated for postnatal depression, and who wind up in plastic surgeons' offices to have as many possible shreds of evidence of this extreme physical process erased.

But what if it didn't have to be that way? I think we've pedestalised motherhood historically (without necessarily supporting mothers themselves), the way we tend to revere anything that looks very challenging or difficult. It's an impressive feat and women seem to take it in their stride. Therefore there must be something mystical, magical and exalted about it. I think this is faulty reasoning.

I don't deny that there are parts of the experience of pregnancy, breastfeeding and giving birth that must subjectively feel magical. To some degree women are literally getting high from a heady cocktail of hormones and neurotransmitters. But it doesn't follow that this is the best way of doing things, or the most privileged position to be in. I think it's very clear that fathers get the better end of the deal on the biological front and that giving women the *option* of becoming mothers in a father-like way would be a great thing.

The main reason we don't discuss this, or hear people loudly advocating for it, is because we don't believe it's possible. Instead, modern debates over the challenges of modern motherhood and parenting seem to stagnate at the question: how can we change our values and social institutions to support women more and make this situation better? While that's an important interim question, I'm always surprised that nobody takes it any further and asks: how can we change this biological imperative so that it no longer puts the sexes in constant tension and forces women to either choose between work and family (or between *person*hood and motherhood), or juggle them and feel like a failure in both domains?

Sometimes it pays to lie to yourself

As I was writing this chapter, quotes from male artists and thinkers kept springing to mind. Why, I wondered, did many of the most honest reflections about the trade-offs incarnate in female bodily design seem to come from men? In 1972, the American writer, infantryman, physics instructor and life-extensionist, Robert Ettinger, wrote that:

> Attempts to make women 'equal' have seemed hopeless (as well as indecent) to some in view of apparently changeless biological function – carrying children in the womb, child-bearing, and suckling. These represent physical handicaps which have given women nearly slave status in most cultures; but they also account for her elevation to a position of acknowledged nobility and special claims. A father is just a man, but a mother is a martyred saint. This situation, however, seems near an end.

Ettinger believed that breastfeeding 'must go because in too many ways it degrades the woman'. He opined that:

It reduces her to a biological machine, an elemental function rather than a fully human person. It restricts her, physically and psychologically; it sets narrow limits on things she can do and times she can do them. It interferes with her career. It may alter her sex life. In short, it is an intolerable imposition, which was once a virtue only through necessity.[14]

It took me a while to figure out why men, and child-free women, seemed to be among the most honest voices exploring the limitations that women's reproductive functions impose on them. Then it struck me. It has everything to do with skin in the game. Motherhood has always been a bad deal for women. Until very recently it was a great way to get yourself killed. But it's been the only real deal available, barring spinsterhood or a nunnery. And in many ways, that's still true.

Like it or not, my fellow XX inheritors, your womb and ovarian reserves are massive stores of value. Yet the design of this currency system is a cruel one. You were born with all your eggs, and they've been ageing and declining in quality and numbers throughout your life, to the point where you're functionally barren by middle age. You attain serious evolutionary value from the day you start menstruating – which eerily commences today around the age of 11, when you're still an emotional and intellectual child. Your best reproductive years are behind you by age 30 – right around the time you're becoming an interesting, experienced, sexually mature person – and by 40 the game is essentially up.

The patriarchy didn't design these cruel fetters, evolution did. Our value as human beings is so deeply bound up with our sexuality and that's a source of tremendous dissatisfaction for so many women. So why aren't more women calling bullshit on this biological design? Because there is nothing else that provides us with such a store of value in the world, however ephemeral. Our uniquely female reproductive functions make us different, special, irreplaceable. We can do something men can't – and because of that, we can get men to do things for us, and we're programmed to like it when men do things for us. We like it when they compete over us. We like it when they strive to impress us, invest time and money in us, and engage in displays of dedication.

We've all heard a woman proudly proclaim somewhere on social media that she has the most important job in the world. 'I literally made another person this year, what did *you* do?' This is a fair point to make when arguing for women's unpaid work to be valued more. I support that cause wholeheartedly for the remaining decades where it's likely to remain a relevant one in human societies. But this quip gets used in another way too. It's a signal of intrinsic value, tied to female anatomy and feminine identity. I'm special. Look what I can do. You *need* me, now show some respect.

But here's the thing about this special function. You're the desktop printer, not the engineer at Hewlett-Packard. You're an assemblage of generic components, put together in a tried and tested way. You blinked and found yourself existing with the capability to incubate life, the way a printer is unboxed, plugged in and finds that it can produce pages of text and imagery of somebody else's design. Let's be frank, if a teenager with an IQ of 75 can perform this 'miracle', and billions of women share this capability, are women *really* performing a miracle? Do we as individuals contribute anything brilliant, ingenious or remarkable to this act of cellular assembly? Or are we just crude meatsack vehicles attempting to take credit for evolution's clever feat of engineering?

I think men, and child-free women, are more open to the view that it might be the latter. But it's much harder for the overwhelming majority of women who become mothers to see it that way, and understandably so. These reproductive functions are the defining aspects of nature's ingrained human life script for a female. What would we do without the script? What would we live for? What would be the point of us? Who would elevate us to the position of martyred saint if we stopped acting as walking human incubators? And so we say, 'No, no, no, motherhood is beautiful. Keep me on my pedestal, I don't want to fall'.

To which my rejoinder is: why not rise instead? I support equality of opportunity for the sexes under the law, but I also second Germaine Greer's lifelong stance against 'equality feminism', by which she means equality of outcome. There is no such thing as true equality of the sexes, because men and women achieve fulfilment in overlapping, but divergent ways. We employ different reproductive strategies under different environmental pressures, and occupy overlapping but divergent social niches. In short, we are similar, but we are not the same.

If women truly want to be more than a set of elemental functions, then it might be time to let go of our uniquely female stores of value when the opportunity arises. But letting go of the purpose, identity and (in a darker sense) insurance against total subjugation, or annihilation, that we've held for millions of years, will not be easy. I would not advocate for it in a world without life extension and human-level artificial intelligence. But if we get there, and move away from forms of personhood circumscribed by sex, men and women alike might finally be free.

Another awkward transition

Like vinyl records, CDs and other technologies that have slowly faded rather than becoming anachronisms overnight, it will be a while before children disappear completely. But this is the first century where it's possible that human fertility rates will plummet to near zero virtually overnight without triggering, or stemming from, societal collapse. All it would take is for the current young generation of breeders to stop believing in 20th-century life-scripts and be sucked deep into the world of soma.

I don't think procreation will fall off the cliff with Gen Z, though I wager fertility rates will keep declining. But enough members of that generation will probably stretch themselves every which way to live what they were raised to believe a 'normal' human life looks like. Yet I think the disruptive cataclysm of the end of having babies could plausibly be realised by the generation that succeeds them. That is, the babies, toddlers and glints in the milkman's eyes of today.

But what if I'm wrong about this being a transhuman century? Suppose life in a hundred years looks just the same as it does today. In that case, even I think it makes sense for most people to have children. I'd probably do it myself. *What else is there?* I mean, there's stuff … you can travel, write books, meet interesting people, but none of that scratches the great existential itch, brings us back down to earth, grounds and fixes us towards a clear and manageable purpose. Very few people can go and create that feeling of purpose as simply or effectively as biology can by turning us into parents. The old script, though imperfect, works.

So, should you have kids today? For many people, I think the answer is still yes. But with the caveat that for just about everybody who undertakes this very serious and lifelong commitment, it will be harder to pull off than it used to be, and will feel like more of a strain, which is oddly paradoxical in a world of such abundance.

I think many parents-to-be will undergo a strange experience as they slowly come to the realisation, as their kids grow up, that they're among the last human generations for whom this practice of childbirth and child-rearing is seen as normal or desirable. It may prove very difficult to watch the world pull in a more cyborgian direction, while you remain more firmly tethered to the old life-scripts. There will also be a need to talk to kids at some point about the crazy transitions happening in their time. The children's books we read to them when they're little won't give them any frame of reference for what's coming. Maybe somebody should write *Spot's Robot Adventure*?

Meanwhile, for a greater proportion of people than ever, the answer to the question 'Should I have kids?' is probably no. Remember it's not just about your immediate desire, it's about your ability to nurture and support this person for twenty years or more. Do you have the emotional and financial toolkit not to make a mess of it? Or are you still a kid yourself? For the growing proportion of young adults who still resemble children, I think it will be sexbots, soma and safe spaces that define their potentially indefinite period of kidulthood, not parenthood.

The crucial thing to remember is that the end of having babies doesn't have to be the end of all that is good and beautiful in life. It's hard to believe that now because we can't imagine a future where there is something better than sex, and romantic and familial love.

Where will the beings of tomorrow come from, if not from our loins?

They'll come from our minds.

What could such a future look like? How could it possibly be good? That is what I am asking you to dare to start imagining.

POSTSCRIPT

THE START OF
SOMETHING NEW

*Whoever makes a shelter of reeds and hides has joined his
spirit to the common destiny of creatures and he will subside
back into the primal mud with scarcely a cry. But who
builds in stone seeks to alter the structure of the universe
and so it was with these masons.*

Cormac McCarthy, *Blood Meridian*

*Nature, Mr Allnut, is what we are put on this earth to rise
above.*

Katharine Hepburn as Rose Sayer in *The African Queen*

It's a funny feeling when it strikes you that *you* might be the
collateral damage of history. That should only be something
that can happen to the people of the past, like those poor saps
who were mown down by machine-gun nests, or impaled
by spears and bayonets in pointless wars. Or the victims of
preventable diseases like polio and smallpox. They seem like
quaint characters in an old black and white film. Their plight
is moving and compelling, but it's hard to imagine that they
were ever fully real.

Yet you and I are no different, but for the fact that we are still alive, and that future generations have yet to look back on us with familial fondness, pity, and that same detached feeling we harbour for historical characters who seem to belong more to a fable than to the world. Societies, technologies and civilisations march on, and as they do, the people of the past begin to look simpler and cruder, resembling pawns controlled by larger forces, more than beings with power and agency.

If you're reading this, you've made it through a book that was designed to question sacred cows – the untouchable, immutable constants of our reality, from ageing and death to childbirth and embodiment. It might have felt like I was knocking them down like skittles in a cluster, so I apologise if you feel a little bowled over. I'm not trying to topple *you*. But if you aren't a little discombobulated, then I haven't done my job. My aim was to rattle your mind's cage with some new ideas that I think could help prepare you as life on spaceship Earth grows stranger, in this time of accelerating change.

To anyone who found the ideas in this book scary, confronting or upsetting, I understand. Many of them are, especially if you're hearing them for the first time. And if you've spent your life and career focusing on humans, humanism and the humanities, it makes sense that a new paradigm heralding a posthuman era sounds like a threat to all you have spent a lifetime exploring, revering and promoting. But I'm afraid the human era *is* coming to an end and we must start thinking about how to navigate this transition and how, or if, we can survive it.

A superhuman future is not the inevitable next step in human evolution. But it *is* humanity's best shot for achieving a sustainable future. As long as we remain human and earthbound, we are sitting ducks who are extremely vulnerable to climate change, natural disasters, pathogens and the ever more powerful arsenal of our own technologies that we are too ape-brained to wield safely in perpetuity.

A posthuman future is not guaranteed to be positive, and there are many possible versions of a posthuman future that could be bad, or subjectively bad for us if they're not conducive to our own survival as human or upgraded beings. But there are no versions of a human future that are sustainable in perpetuity. And humanity will be hard pressed to make it through this century alive *if we do not* invent technologies that render us less susceptible to short-term thinking, cognitive biases, tribalism and human stupidity.

In the final chapter of this book I asked you to dare to imagine a posthuman future – a world where no babies are born, yet one that's better than the reality we know and love. I know that's a mind-bendingly hard concept to fathom, and many of you might have come to the end of that chapter and thought the suggestion preposterous. I want to make one final attempt in these closing pages to show you that it's possible.

The view from posthumanity

Imagine, for a moment, that you're a cat. You have the brain of a cat, the programming and motivations of a cat, and let's just ignore the incongruous fact that you can understand a human thought experiment. What does your life as a cat entail? You live what we might call a simple, and very instinctive kind of life. You have a hunting instinct, you learn through play, you mate, you chase mice and birds, and you kill other animals for sport or sustenance when the opportunity arises.

There's not much else to 'catmanity'. There might be some nice colourful yarn to play with, but there are no Picasso paintings to behold. There's no internet, no horticulture, no philosophy, and no conception of the fact that there is a universe, planets, stars, matter or DNA. There is not even an inkling that these things exist, or that they *could* exist. As a cat, you do not have enough brain capacity to conceive of them. And because of that, it's impossible for you to comprehend their value to a species that can imagine them.

A cat can probably think of no greater pleasure than chasing after a bird and devouring it. It would violate the very essence of cathood to say, hang on, wouldn't it be better if we figured out a way to stop wanting to kill other animals for sustenance and pleasure? Aren't there higher sets of values and broader spectrums of experience? Should we maybe look into that at some point? No, replies the cat's adamant programming, because we would lose our very catmanity in the process. The instinct to hunt and kill is part of what makes us who we are.

And the cat – let's call him Felix – is right. Tethered to the brain, body and evolutionary design of a cat, it's impossible to imagine anything better, higher or even just very different. Big changes beyond the cat's known realm of experience could not be imagined, including this thought experiment. If they could, they would probably be viewed as a loss of self, and an annihilation of the source of all known motivation, purpose and fulfilment.

But you *can* conceive of more than catmanity. Unlike Felix, you know what it's like to be moved to tears by a beautiful piece of music, and to have the thoughts of another being beamed into your brain as you read words on a page, allowing minds to commune across the ages, long after the originators of the thoughts have died. You know what it is to sit in front of a delicious piece of steak and think, I wonder what the life of this cow was like, and should I really be eating this? So, if I gave you the choice to be a cat or a human, which would you choose?

Somehow I don't think many of you would choose to be a cat. Humans like being human. We think it's incredible that we can do all sorts of things that no other species has ever done, felt or thought and we're very unwilling to give any of this humanity up. It's wondrous and awe-inspiring and it's all we've ever known. But it's still worth asking: *is this the best possible mode of existence?* Do we have access to the deepest, richest layers of experience, the most knowledge that it's possible to garner, the highest peaks of pleasure? Are we able to imagine the most thrilling kinds of intellectual challenges? Or something better and more satisfying than sex, and love? Can we really imagine the subjective experience of inhabiting a mind that is orders of magnitude more powerful and capable than our own? I think it's very clear that the answer is no.

But why do we need to aspire to such things? Is more power and more intelligence inherently better? Isn't there something beautiful about being simple and flawed, and embracing what you are? That's exactly what the cat would say. And he's not totally wrong, it's a valid subjective thought. But Felix also can't see the larger picture and he has no idea what he's missing. We're the same when it comes to beings more capable than ourselves. Humans are to future beings what cats are to humans: simpler, cruder, and unable to imagine how much more there can be to life.

A superhuman being who had much more brain capacity than us would have a very hard time trying to explain to us what it's like to be them, and how expansive and enriching some of their experiences are. It would be like us trying to explain to Felix why we love Mozart, or Kanye West.

You see Felix, there's this thing called music. It's like sounds, but special sounds arranged in these amazing ways, and when you hear them it changes how you feel. You'd love it if you could experience it. Trust us, it would make your life so much richer.

But even if Felix could understand the words we were saying, the logic wouldn't make any sense. Felix knows what sounds are. He hears sounds all the time. They spook him sometimes and he's got better hearing than we do. But why would he want *more* sounds? What would he do with those sounds? Without experiencing them as a human does, it's hard to believe they're of any value at all.

They might even appear to be of negative value, because if Felix wasted his time listening to sounds all day, he might not chase any birds or do any cat things, and cat things are the important things. Of course that's exactly how we think about human things. As humans who only consider value and fulfilment from a human perspective, we don't consider whether we're pursuing the *best* possible way of doing things. We're more likely to proceed with something because that's the way we've always done it and worry that if we stopped doing things the usual way we might lose an important part of ourselves. Yet we hardly ever think about how much we might stand to gain.

The only way out is up

Despite the social turmoil and future shock that we're starting to experience, the transhuman era is a transitional stage that humanity must pass through. We're like passengers and crew flying through a storm over the equator, the jet being buffeted by unpredictable jolts of turbulence, rattling and occasionally plummeting a few hundred feet before climbing again.

The storm surrounds us on all sides. We can't fly over or under it, there's nowhere to make an emergency landing, and we don't have enough fuel to get back to where we started. Sure, we could panic, assume the voyage is doomed and start fighting over the remaining packets of peanuts. Or we could stay calm, keep flying, and pay extra attention to altitude, wind resistance and weather, steering our way through the worst of the turbulence as best we can.

If we fly on autopilot and fail to pay attention to bleeping sensors and danger warnings, the aircraft could stall and fall from the sky. But this is not the most likely outcome if we pay attention. We have the tools, knowledge and machinery to be able to withstand most kinds of storms. For the best shot at a good outcome, we need to be closely attentive to the conditions of the immediate present. But we must also keep a close eye on the horizon, chart a course for clearer skies, and look for a safe place to land, so we can refuel, take stock, and continue our adventures.

Embracing posthumanity is a new way to think. It might sound scary, but you can't afford to ignore the ideas in this book because the transhuman era is already in full swing. Challenging yourself to think of humanity the way we view other species, like cats, can feel uncomfortable. But

I think the English philosopher John Stuart Mill was on to something when he famously quipped:

> It is better to be a human dissatisfied than a pig satisfied. Better to be Socrates dissatisfied than a fool satisfied. And if the fool or the pig are of a different opinion, it is because they only know their side of the question.[1]

I've often questioned this sentiment and sometimes think it might be better to be a 'fool'. But put two levers in front of me and stick a gun to my head and I know I'll pull the one marked Socrates. A world of satisfied fools has little hope of durability. In the long run the same is true of a world of humans engrossed in our equivalents of chasing birds and playing with colourful yarn. Pigs, fools and humans have no choice but to accept what they cannot change. Superhumans could change what is unacceptable about the world and pave the way for more radiant beings to enjoy the wonders of existence for millions of years to come. For all the inherent dangers, the race is on to invent them before the doomsday clock strikes twelve.

ACKNOWLEDGMENTS

I'm enormously grateful to the many friends and colleagues who offered feedback at various stages of this project.

Some did rigorous reads of the final manuscript, while others let me try out bits and pieces on them earlier in the process. Friends and family members were also invaluable sounding boards, allowing me to bounce ideas around with them during our epic, multi-hour phone natters.

Huge thanks, in alphabetical order, to: Angela Aristizábal, Spencer Becker-Kahn, Damon Binder, Russell Blackford, Dominic Bohan, Mirella Bohan, Tom Bohan, Rob Brooks, David Christian, Max Daniel, Michael Gillings, David Goldie, Niko Iliakas, Tushant Jha, Malin Kankanamge, Jessica Kirkness, Neil Levy, Greg Lewis, Alex Lintz, Toby Newberry, Anders Sandberg and Hugh Spillekom.

The superb cover design for *Future Superhuman* is the brain-child of my art-school chum turned graphic designer Arielle Nguyen. Many thanks for taking the project on at such a busy time.

A final hat tip goes to the many spectacular humans in my life in 2020–2021. Thank you. I love you all.

NOTES

Preface

1   JBS Haldane, *Daedalus: or Science and the Future*, EP Dutton & Company, New York, 1923, p. vii.

2   HG Wells, *World Brain*, Methuen & Co Ltd, London, 1938, p. 40.

3   Wells, *World Brain*, p. 2.

Introduction: The big picture and the human story

1   See, for example: Andy Clark, *Natural Born Cyborgs*, Oxford University Press, Oxford, 2003.

1   Prepare for the future shock

1   Tim Urban, 'The AI revolution: The road to superintelligence', Wait But Why, 22 January 2015, <waitbutwhy.com/2015/01/artificial-intelligence-revolution-1.html>.

2   The basic concept of inoculation through variolation can be traced back to at least 16th-century China, and possibly back to 1000 AD, though evidence for such early origins are scant. See: Alexandra Flemming, 'The origins of vaccination', *Nature*, 28 September 2020, <www.nature.com/articles/d42859-020-00006-7>; Arthur Boylston, 'The origins of innoculation', *Journal of the Royal Society of Medicine*, vol. 105, no. 7, 2012, <www.ncbi.nlm.nih.gov/pmc/articles/PMC3407399/>. This predates Edward Jenner famously vaccinating an 8-year-old boy against smallpox with cowpox vesicle fluid in 1796. The idea of variolation was imported to Europe from Istanbul in the early 18th century, and was the subject of increasing interest and acceptance among the aristocracy from the 1720s onwards. See: Stefan Riedel, 'Edward Jenner and the history of smallpox and vaccination', *Baylor University University Medical Center Proceedings*, vol. 18, no. 1, 2005, <www.ncbi.nlm.nih.gov/pmc/articles/PMC1200696/>. However, modern vaccines would still have been considered a medical marvel in 1750. Hypodermic needles weren't invented until the 1840s, and the idea of routine childhood vaccination and mass vaccination programs for all manner of infectious diseases would have been mind-blowing, particularly when viewed alongside actuarial tables demonstrating their effects on child mortality and life expectancy. Even the experience of being jabbed with a sterile needle would have been totally alien to people of this time.

3   Lest any readers worry that I'm imagining everyone was dead by their fortieth birthday in the Palaeolithic, rest assured I know that was not

the case. Average life expectancy was dragged down by very high rates of infant mortality, but if you survived childhood, you had a reasonable chance of living past forty. Nevertheless, the proportion of people now living past forty, and well into their sixties, seventies and eighties is much greater today than at any other point in human history, both in the UK and globally.

4   Steffen and colleagues have compiled a series of Great Acceleration charts. I've used some of their data in my 'Global trends since the industrial revolution' chart. I've used Steffen's term the 'Great Acceleration' to describe the overall phenomenon of rapid, global, post-industrial transformation because it's a helpful epithet. However, I've included some different metrics, which I think highlight other crucial parts of the same story, such as the global literacy rate and the percentage of humans not living in extreme poverty.

5   Population data sourced from: Our World in Data, 'Total world population – Comparison of different sources', <ourworldindata. org/grapher/total-world-population-comparison-of-different-sources?>; International Geosphere-Biosphere Programme (IGBP), 'Great Accleleration Data,' 2015, <www.igbp.net/globalchange/greatacceleration.4.1b8ae20512db692f2a680001630.html>. Average life expectancy data sourced from: Our World in Data, 'Life-expectancy at birth, including the UN projections', 2020, <ourworldindata.org/grapher/life-expectancy-at-birth-including-the-un-projections?>; Luke Muehlhauser, 'How big a deal was the Industrial Revolution?', personal website, no date, <lukemuehlhauser. com/industrial-revolution/>. Real GDP data sourced from: IGBP 'Great Acceleration Data'. Percentage not living in extreme poverty data sourced from: Luke Muehlhauser, 'How big a deal was the Industrial Revolution?'. Literacy rate data sourced from: Our World in Data, 'Literate and illiterate world population', no date, <ourworldindata.org/grapher/literate-and-illiterate-world-population_2?country=~OWID_WRL>. Percentage living in a democracy sourced from: Our World in Data, 'World citizens living under different political regimes', 2016, <ourworldindata.org/grapher/world-pop-by-political-regime?country=~OWID_WRL>; Luke Muehlhauser, 'How big a deal was the Industrial Revolution?'. Primary energy use data sourced from: IGBP, 'Great Acceleration Data'. Energy capture data sourced from: Luke Muehlhauser, 'How big a deal was the Industrial Revolution?'. Internet users data sourced from: Our World in Data, 'Share of the population using the internet, 1990 to 2019', 2019 <ourworldindata.org/grapher/share-of-individuals-using-the-internet?tab=chart>. War-making capacity data sourced from: Luke Muehlhauser, 'How big a deal was the Industrial Revolution?'. Carbon dioxide data sourced from: IGBP 'Great Acceleration Data'. Biodiversity loss data sourced from: IGBP 'Great Acceleration Data'.

6   'Children: Improving survival and wellbeing', WHO, 2020, <www.
     who.int/news-room/fact-sheets/detail/children-reducing-mortality>.
7   Ray Kurzweil, *The Singularity Is Near: When Humans Transcend Biology*,
     Penguin, New York, 2006. Urban, 'The AI revolution: The road to
     superintelligence'.
8   For a more rigorous exploration of this possibility, see: Robin Hanson,
     'Long-term growth as a sequence of exponential modes', George
     Mason University, 2000, <mason.gmu.edu/~rhanson/longgrow.pdf>.
9   There's plenty of literature suggesting that Moore's Law is slowing
     down and that we're resorting to these techniques as work-arounds
     to squeeze the last little bit of progress out of a stagnating trend.
     There does not appear to be consensus on how much steam is left
     in the Moore's Law engine of progress. What's clear is that we're
     nowhere near approaching the physical limits of computation. Nor
     can we be certain of how promising emerging paradigms like quantum
     computing may prove in the 21st century. What we do know is that
     there are many avenues by which we can theoretically harness more
     computing power. The more avenues we have available, the higher the
     probability that some of them will succeed.
10   Kurzweil, *The Singularity Is Near*.
11   'Apple II Personal Computer', National Museum of American
     History, no date, <americanhistory.si.edu/collections/search/object/
     nmah_334638>.
12   Alex Crippen, 'Warren Buffett buys this with his billions and it makes
     him happy', CNBC, 12 November 2012, <www.cnbc.com/2012/11/12/
     warren-buffett-buys-this-with-his-billions-and-it-makes-him-happy.
     html>.
13   David Rotman, 'We're not prepared for the end of Moore's Law',
     MIT Technology Review, 24 February 2020,.
14   James Lovelock, *Novacene: The coming age of hyperintelligence*, Penguin,
     Kindle, 2020.
15   In their survey of four expert groups in 2012 and 2013, Vincent C
     Müller and Nick Bostrom report, 'The median estimate of respondents
     was for a one in two chance that high-level machine intelligence will
     be developed around 2040–2050, rising to a nine in ten chance by
     2075. Experts expect that systems will move on to superintelligence
     in less than 30 years thereafter'. See: Vincent C Müller and Nick
     Bostrom, 'Future progress in artificial intelligence: A survey of expert
     opinion' in Vincent Müller (ed.), *Fundamental Issues of Artificial
     Superintelligence*, Springer, 2016, pp. 553–571, <philpapers.org/rec/
     MLLFPI>. In a 2015 survey of machine learning researchers, Katja
     Grace et al report, 'the aggregate forecast gave a 50 per cent chance
     of HLMI [high-level machine intelligence, which can perform most

tasks as well or better than humans unaided] occurring within 45 years and a 10 per cent chance of it occurring within 9 years'. Notably, 'Asian respondents expect HLMI in 30 years, whereas North Americans expect it in 74 years'. See: Katja Grace et al, 'Viewpoint: When will AI exceed human performance? Evidence from AI experts', *Journal of Artificial Intelligence Research*, vol. 62, 2018, pp. 729–754, <arxiv.org/abs/1705.08807>. In Toby Walsh's 2017 survey, the median prediction for a 10 per cent probability of HLMI being developed was 2033 for robotics experts and 2035 for AI experts. For a 50 per cent probability it was 2065 and 2061 respectively. And for a 90 per cent probability it was 2118 and 2109. See Toby Walsh, 'Expert and non-expert opinion about technological unemployment', *International Journal of Automation and Computing*, vol. 15, no. 5, October 2018, pp. 633–638, <https://doi.org/10.1007/s11633-018-1127-x>.

16 Rory Cellan-Jones, 'Stephen Hawking warns artificial intelligence could end mankind', BBC, 2 December 2014, <www.bbc.com/news/technology-30290540>.

2   T is for transhumanism

1   Larry McCaffery, 'An interview with David Foster Wallace,' *Review of Contemporary Fiction*, Summer 1993, <samizdat.co/shelf/documents/2005/03.07-dfwinterview/dfwinterview.pdf>.

2   *The Joe Rogan Experience* podcast, '#1470 – Elon Musk', May 2020.

3   Neuralink 2020 progress update, <www.youtube.com/watch?v=DVvmgjBL74w>.

4   'Who we are', Methuselah Foundation, <www.mfoundation.org/who-we-are#about-us>.

5   George Packer, *The Unwinding: An inner history of the New America*, Farrar, Straus and Giroux, New York, 2013.

6   Ariana Eunjung Cha, 'Peter Thiel's quest to find the key to eternal life', the *Washington Post*, 3 April 2015, <www.washingtonpost.com/business/on-leadership/peter-thiels-life-goal-to-extend-our-time-on-this-earth/2015/04/03/b7a1779c-4814-11e4-891d-713f052086a0_story.html>.

7   Peter Thiel, 'Foreword: The problem of death' in Sonia Arrison, *100 Plus: How the coming age of longevity will change everything, from careers and relationships to family and faith*, Basic Books, New York, 2011.

8   Chan Zuckerberg Initiative, 'Supporting scientific research to cure, prevent, or manage all diseases in our children's lifetime', 21 September 2016, <www.chanzuckerberg.com/newsroom/supporting-scientific-research-to-cure-prevent-or-manage-all-diseases-in-our-childrens-lifetime>.

9   Mark Zuckerberg, town hall Q&A, Facebook, 1 July 2015, <www.facebook.com/zuck/posts/10102213601037571?_rdr=p&pnref=story>.

10 Hans Moravec, *Mind Children: The future of robot and human intelligence*, Harvard University Press, Cambridge, MA, 1988, p. 15.

11 Tom Simonite, 'Google's intelligence designer', *MIT Technology Review*, 2 December 2014, <www.technologyreview.com/s/532876/googles-intelligence-designer/>.

12 'About', DeepMind, <deepmind.com/about/>.

13 Dawn Chan, 'The immortality upgrade', the *New Yorker*, 20 April 2016.

14 Chase Peterson-Withorn, 'From Rockefeller to Ford, see Forbes' 1918 ranking of the richest people in America', *Forbes*, 19 September 2017, <www.forbes.com/sites/chasewithorn/2017/09/19/the-first-forbes-list-see-who-the-richest-americans-were-in-1918/#53b7bbc94c0d>.

15 Both of these figures have been converted from 1918 to 2017 US dollars. See: Peterson-Withorn, 'From Rockefeller to Ford'.

16 *Forbes*, 'The real-time billionaires list', <www.forbes.com/real-time-billionaires/#5c18c6c43d78>.

17 The net worth of the following people was added to reach a combined total of $623 billion in February 2021: Jeff Bezos, Elon Musk, Bill Gates, Mark Zuckerberg, Warren Buffett, Larry Page, Sergey Brin, Larry Ellison, Steve Ballmer, and Michael Dell. For real-time updates, see: *Forbes*, 'The real-time billionaires list'.

18 Jeff Bercovici, 'Peter Thiel is very, very interested in young people's blood', *Inc.*, no date, <www.inc.com/jeff-bercovici/peter-thiel-young-blood.html>; Sean Keach, 'Tech Tikes', the *Sun*, 16 July 2019, <www.thesun.co.uk/tech/9515905/silicon-valley-transhumanism-cyborg-robot-augment/>.

19 Bret Weinstein, 'A shared frontier – a letter from Bret', personal website, no date, <bretweinstein.net/>.

20 David Sinclair and Matthew D LaPlante, *Lifespan: The revolutionary science of why we age – and why we don't have to*, Thorsons/HarperCollins, London, Kindle, 2019.

21 Martin Rees, 'Organic intelligence has no long-term future', *Edge*, 2015, <www.edge.org/response-detail/26160>.

22 Jim AC Everett, Nadira S Faber, Julian Savulescu & Molly J Crockett, 'The costs of being consequentialist: Social inference from instrumental harm and impartial beneficence', *Journal of Experimental Social Psychology*, November 2018, <www.ncbi.nlm.nih.gov/pmc/articles/PMC6185873/>.

23 Mark O'Connell, *To Be a Machine: Adventures among cyborgs, utopians, hackers, and the futurists solving the modest problem of death*, Granta, London, Kindle, 2017.

24 For the history of transhumanist demographics, see: Elise Bohan, *A History of Transhumanism*, doctoral thesis submitted for examination, Macquarie University, 2018. Macquarie University Research Online, <hdl.handle.net/1959.14/1271515>.

3   What's so great about humanity anyway?

1   Richard Dawkins, *Unweaving the Rainbow: Science, delusion and the appetite for wonder*, Houghton Mifflin, Boston, 1998, p. 1.

2   'Number of animals slaughtered for meat, World, 1961 to 2018', Our World in Data, no date, <ourworldindata.org/grapher/animals-slaughtered-for-meat>.

3   Mia Fernyhough, Christine J Nicol, Teun van de Braak, Michael J Toscano & Morten Tønnessen, 'The ethics of laying hen genetics', *Journal of Agricultural and Environmental Ethics*, vol. 33, 2020, pp. 15–36, <link.springer.com/article/10.1007/s10806-019-09810-2>.

4   William Ralph Inge, *Outspoken Essays, Second Series*, Longmans, Green and Co., London, 1922, p. 183.

5   Australian women spend $15 billion per year on grooming, while Australian men spend $7 billion. See: Georgina Dent, 'Australian women pay a high price for looking good', *Sydney Morning Herald*, 21 July 2017, <www.smh.com.au/money/australian-women-pay-a-high-price-for-looking-good-20170720-gxfcxi.html>. Female American Millennials spend 42 per cent more than their male counterparts on apparel and personal care. See: Stephanie Horan, 'How millennial men and women spend money', *Smartasset*, 12 February 2020, <smartasset.com/checking-account/how-millennial-men-and-women-spend-money-2020>. Estimates of monetary expenditure vary across countries and looks maintenance categories (for example, clothes, shoes, toiletries, skin products, manicures, haircare, gyms, personal trainers and cosmetic surgery), but double the spend is a reasonable aggregate estimate. Globally, women also spend more time each day on personal appearance and grooming. See: M Ridder, 'Estimated daily time spent on personal appearance and grooming worldwide as of 2016, by gender', Statista, 23 November 2020, <www.statista.com/statistics/805935/daily-time-spent-on-personal-appearance-and-grooming-by-gender-worldwide/>.

6   Hava Tirosh-Samuelson, 'Engaging transhumanism', in Gregory R Hansell and William Grassie (eds), *H+/-: Transhumanism and Its Critics*, Metanexus Institute, Philadelphia, 2011.

7   John Keats, 'To Benjamin Bailey, 10 June 1818', in Robert Gittings (ed.), *John Keats: Selected Letters*, Oxford University Press, Oxford, 2002, p. 94.

8   Jenna Ross, 'Global deaths: This is how COVID-19 compares to other diseases', World Economic Forum, 16 May 2020, <www.weforum.org/agenda/2020/05/how-many-people-die-each-day-covid-19-coronavirus/>; Britt Wray, 'The ambitious quest to cure ageing like a disease', BBC, 5 February 2018, <www.bbc.com/future/article/20180203-the-ambitious-quest-to-cure-ageing-like-a-disease>.

9   Ricki Harris, 'Elon Musk: Humanity is a kind of "biological boot loader" for AI', *Wired*, 1 September 2019, <www.wired.com/story/elon-musk-humanity-biological-boot-loader-ai/>.

10   'The transhumanist FAQ', Nick Bostrom, p. 30, <www.nickbostrom.com/views/transhumanist.pdf>.

4   The genetic lottery

1   Tim Adams, 'The stuff of life', the *Observer*, 6 April 2003, <www.theguardian.com/education/2003/apr/06/highereducation.uk1>.

2   David A Frederick and Brooke N Jenkins, 'Height and body mass on the mating market: Associations with number of sex partners and extra-pair sex among heterosexual men and women aged 18–65', *Evolutionary Psychology*, 18 September 2015, <https://doi.org/10.1177/1474704915604563>.

3   See: David A Frederick and Brooke N Jenkins, 'Height and body mass on the mating market'; Eunice Y Chen and Molly Brown, 'Obesity stigma in sexual relationships', *Obesity Research*, vol. 13, no. 8, 2005, <https://doi.org/10.1038/oby.2005.168>; Rebecca Puhl and Dr Kelly D Brownell, 'Bias, discrimination, and obesity', *Obesity Research*, vol. 9, no. 12, 2012, <https://doi.org/10.1038/oby.2001.108>; Renate van der Zee, 'Demoted or dismissed because of your weight? The reality of the size ceiling', the *Guardian*, 30 August 2017, <www.theguardian.com/inequality/2017/aug/30/demoted-dismissed-weight-size-ceiling-work-discrimination>.

4   See: Steve Stewart Williams, 'Survival of the prettiest' in *The Ape That Understood the Universe: How mind and culture evolve*, Cambridge University Press, Cambridge, 2018.

5   Alex Duval Smith, 'Girls being force-fed for marriage as fattening farms revived', the *Observer*, 1 March 2009, <www.theguardian.com/world/2009/mar/01/mauritania-force-feeding-marriage>; Abigail Haworth, 'Forced to be fat', *Marie Claire*, 21 July 2011, <www.marieclaire.com/politics/news/a3513/forcefeeding-in-mauritania/>.

6   I'm not arguing that cultural norms can't be shifted at all. Greater appreciation for people of a range of sizes and appearances can be instilled. I'm specifically arguing about asymmetry and *morbid* obesity here. Even if an outlier culture celebrated morbid obesity as a status signal, that preference is unlikely to become universal cross-culturally, or persistent within that culture over very long periods of time.

7   See: Steve Stewart Williams, 'Maximising returns on parental investment' in *The Ape That Understood the Universe*.

8   See: Daniel Nettle, 'Women's height, reproductive success and the evolution of sexual dimorphism in modern humans', *Proceedings of the Royal Society B*, 22 September 2002, <https:doi.org/10.1098/rspb.2002.2111>; Christian Rudder, 'The big lies people tell in online dating', *OKTrends*, 7 July 2010, archived at: <web.archive.

org/web/20100814002125/http://blog.okcupid.com/index.php/the-biggest-lies-in-online-dating/>.

9   These latter two examples are deployed here to make the point that 'not all forms of diversity' that arise in the human population are net positive. I am making no claims about how either condition arises, or to what extent they are genetically determined.

10 'Eugenics', 2 July 2014, *Stanford Encyclopedia of Philosophy*, <plato. stanford.edu/entries/eugenics/>.

11 Thirty-two per cent of registered Australian marriages in 2018 were of partners born in different countries. In 2006, the number was only 18 per cent. See: Jason Fang and Christina Zhou, 'Australian migrants share the challenges of intercultural marriages', ABC News, 15 March 2020, <www.abc.net.au/news/2020-03-15/australia-migrants-share-challenges-of-intercultural-marriage/12045598>. In the United States, 17 per cent of newlyweds were intermarried in 2015, an almost fivefold increase since 1967. See: 'Trends and patterns in intermarriage,' Pew Research Center, 18 May, 2017, <www.pewresearch.org/social-trends/2017/05/18/1-trends-and-patterns-in-intermarriage/>. As of 2021, 94 per cent of Americans approve of interracial marriages between black and white Americans, up from 4 per cent in 1958 when Gallup first polled on the question. See: Justin McCarthy, 'U.S. approval of interracial marriage at new high of 94%,' Gallup, 10 September, 2021, <https://news.gallup.com/poll/354638/approval-interracial-marriage-new-high.aspx>.

12 See: Francis Collins, *The Language of Life*, HarperCollins, Kindle, 2010.

13 The science is complex. For an excellent recent account of the predictive powers of genomic sequencing and the polygenic scores, see: Robert Plomin, *Blueprint: How DNA makes us who we are*, Penguin Random House, Kindle, 2018.

14 'NHS 100,000 Genomes Project takes centre stage at Healthcare Science Week', National Health Service, 7 March 2016, <www. england.nhs.uk/genomics/100000-genomes-project/>.

15 'The 100,000 Genomes Project', Genomics England, no date, <www.genomicsengland.co.uk/about-genomics-england/the-100000-genomes-project/>.

16 'NIH-funded genome centers to accelerate precision medicine discoveries', National Institute of Health, 25 September 2018, <www.nih.gov/news-events/news-releases/nih-funded-genome-centers-accelerate-precision-medicine-discoveries>.

17 The US Government's National Human Genome Research Institute estimates that the true cost of sequencing the first human genome in 2003 was between $500 million and $1 billion. See: National Human Genome Research Institute, no date, <www.genome.gov/about-genomics/fact-sheets/Sequencing-Human-Genome-cost>. That project took thirteen years, because the scientists involved actually did have

to invent the wheel, so to speak, doing a heck of a lot of preliminary work before they even began sequencing human DNA. With the help of Moore's Law and next-generation sequencing technologies, the cost of sequencing a human genome fell to around $12 million by 2006, then plummeted to $30 000 by 2010, then $4000 in 2015, and is under $1000 today. The true clinical costs are, of course, much higher than what it costs to just get a readout of the data. For one thing, multiple samples usually need to be sequenced for cancer and rare disease diagnoses. But would that matter if the cost of sequencing plummeted even further, to $100, or $10, or practically free? After all, for each megabase of a DNA sequence (or each million bases sequenced) the cost fell from $5000 in 2001, to 35 cents in 2010, to a tenth of a cent in 2019. See: NHGRI, 'DNA sequencing costs', <www.genome.gov/about-genomics/fact-sheets/DNA-Sequencing-Costs-Data>.

18   Tom Diacovo and Gerard Vockley, 'Sequencing the genome of newborns in the US: Are we ready?', the *Conversation*, 28 June 2019, <theconversation.com/sequencing-the-genome-of-newborns-in-the-us-are-we-ready-119091>.

19   James Watson, cited in Tim Adams, 'The stuff of life'. The philosophers Julian Savulescu and Brian Earp explore this idea in their book *Love Drugs: The chemical future of relationships*, Stanford University Press, Redwood City, 2020.

5   Ape-brains in a modern world

1   'Episode #150 – The Map of Misunderstanding: A conversation with Daniel Kahneman', *Making Sense* podcast, 12 March 2019, <www.samharris.org/podcasts/making-sense-episodes/150-map-misunderstanding>.

2   Toby Ord, *The Precipice*, Bloomsbury Publishing, Kindle Edition, 2020.

3   Robert Wiblin, Arden Koehler and Keiran Harris, 'Toby Ord on the precipice and humanity's potential futures', *80,000 Hours* podcast transcript, 7 March 2020, <80000hours.org/podcast/episodes/toby-ord-the-precipice-existential-risk-future-humanity/>.

4   See Gerd Gigerenzer, 'Out of the frying pan into the fire: Behavioral reactions to terrorist attacks', *Risk Analysis*, vol. 26, no. 2, 2006, <web.missouri.edu/~segerti/capstone/OutoftheFryingPan.pdf>; Garrick Blalock, Vrinda Kadiyali and Daniel H Simon, 'Driving fatalities after 9/11: a hidden cost of terrorism', *Applied Economics*, vol. 41, no. 14, 2009, <https://doi.org/10.1080/00036840601069757>.

5   I know the replication crisis took some of the conclusions of forms of this research to task, but it's still a helpful framing tale.

6   Heather Heying on *The Dark Horse* podcast, 'Bret and Heather 1st in a series of Live Stream: Tests, masks, and more', (from 06:27), streamed live on 25 March 2020, <www.youtube.com/watch?v=ym-WGOq96G0>.

7 'Coronavirus: Iran cover-up of deaths revealed by data leak', BBC News, 3 August 2020, <www.bbc.com/news/world-middle-east-53598965>.

8 The *Guardian*, '"I shook hands with everybody," says Boris Johnson weeks before coronavirus diagnosis – video', 27 March 2020, <www.theguardian.com/world/video/2020/mar/27/i-shook-hands-with-everybody-says-boris-johnson-weeks-before-coronavirus-diagnosis-video>.

9 'Bob Woodward book *Rage*: Trump denies lying about risks of coronavirus', BBC News, 10 September 2020, <www.bbc.com/news/world-us-canada-54107677>.

10 Suganya Lakshmi, 'COVID-19 in India: The battle we weren't ready for', *Zafigo*, 22 June 2021, <zafigo.com/stories/20210622-covid-19-in-india/?fbclid=IwAR0esCCvujLnAb3t-43ZedsMxiSuLYh9vCGdK3G KuJfA8XF2Ggaq12ueObo>.

11 Udani Samarasekera, 'India grapples with second wave of COVID-19', the *Lancet*, June 2021, <www.thelancet.com/journals/lanmic/article/PIIS2666-5247(21)00123-3/fulltext>.

12 Douglas Adams, *The Restaurant at the End of the Universe*, Pan Books, London, 2009, p. 176.

6 Unfit custodians of the future

1 For more on the promise and peril of advanced nanotechnology, see: K Eric Drexler, *Engines of Creation*, Anchor Press, Doubleday, 1986; K Eric Drexler, *Radical Abundance: How a revolution in nanotechnology will change civilization*, PublicAffairs, 2013.

2 Shalailah Medhora, 'Coal is the future, insists Tony Abbott as UN calls for action on climate change', the *Guardian*, 4 November 2014, <www.theguardian.com/environment/2014/nov/04/coal-is-the-future-insists-tony-abbott-as-un-calls-for-action-on-climate-change>.

3 James Lovelock, *A Rough Ride to the Future*, Penguin, Kindle, 2014.

4 Hannah Ritchie and Max Roser, 'Renewable energy', Our World in Data, no date, <ourworldindata.org/renewable-energy>.

5 Hannah Ritchie, 'What was the death toll from Fukushima?', Our World in Data, 1 December 2021, <ourworldindata.org/what-was-the-death-toll-from-chernobyl-and-fukushima>.

6 Hannah Ritchie, 'What are the safest and cleanest sources of energy?', Our World in Data, 10 February 2020, <ourworldindata.org/safest-sources-of-energy>.

7 James Conca, 'How deadly is your kilowatt? We rank the killer energy sources', *Forbes*, 10 June 2012, <www.forbes.com/sites/jamesconca/2012/06/10/energys-deathprint-a-price-always-paid/?sh=37ba6119709b>.

8 'Frequently asked questions (FAQs)', US Energy Information Administration (EIA), 9 October 2020, <www.eia.gov/tools/faqs/faq.php?id=97&t=3>.

9   'RBMK reactors – Appendix to nuclear power reactors', World Nuclear
      Association, December 2020, <www.world-nuclear.org/information-
      library/nuclear-fuel-cycle/nuclear-power-reactors/appendices/rbmk-
      reactors.aspx>.

10   Hannah Ritchie and Max Roser, 'Air pollution', Our World in Data,
      October 2017, <ourworldindata.org/air-pollution#air-pollution-is-
      one-of-the-world-s-leading-risk-factors-for-death>.

11   Brad Plumer, 'The rise and fall of nuclear power, in 6 charts', *Vox*,
      30 January 2015, <www.vox.com/2014/8/1/5958943/nuclear-power-
      rise-fall-six-charts>. For relicensing and the lifespan of nuclear plants,
      see: Scott L Montgomery and Thomas Graham Jr, *Seeing the Light:
      The case for nuclear power in the 21st century*, Cambridge University
      Press, Cambridge, Kindle, 2017.

12   Brad Plumer, 'The rise and fall of nuclear power, in 6 charts'; Hannah
      Ritchie and Max Roser, 'Nuclear energy', Our World in Data, no
      date, <ourworldindata.org/nuclear-energy>; 'Nuclear power in the
      world today', World Nuclear Association, <https://world-nuclear.org/
      information-library/current-and-future-generation/nuclear-power-in-
      the-world-today.aspx>.

13   The former Chief Scientific Advisor to the UK Department of Energy
      and Climate Change, the late physicist David MacKay, made these
      points in his 2008 book *Sustainable Energy – Without the Hot Air*, UIT,
      Cambridge, 2009.

14   'Nuclear waste', NEI, no date, <www.nei.org/fundamentals/nuclear-
      waste>.

15   'Exploring the Natrium™ technology's energy storage system',
      TerraPower, 4 November 2020, <www.terrapower.com/exploring-the-
      natrium-energy-storage-system/>; David Grossman, 'TerraPower:
      Why Bill Gates thinks nuclear energy is the future', 6 September
      2020, <www.inverse.com/innovation/bill-gates-thinks-nuclear-energy-
      is-the-future/amp>; 'With Natrium, nuclear can pair perfectly with
      energy storage and renewables', NEI, <www.nei.org/news/2020/
      natrium-nuclear-pairs-renewables-energy-storage>.

16   Greenpeace, 'Nuclear Energy,' 16 March 2021, <web.archive.org/
      web/20210316082320/https://www.greenpeace.org/usa/ending-the-
      climate-crisis/issues/nuclear/>.

17   As Bill Gates explains in his 2021 book, *How to Avoid A Climate
      Disaster: The solutions we have and the breakthroughs we need* (Alfred A.
      Knopf, New York, 2021), 'The wind doesn't always blow and the sun
      doesn't always shine, and we don't have affordable batteries that can
      store city-sized amounts of energy for long enough. Besides, making
      electricity accounts for only 27 percent of all greenhouse gas emissions.
      Even if we had a huge breakthrough in batteries, we would still need to
      get rid of the other 73 percent'.

18   See: Hannah Ritchie, 'Sector by sector: Where do global greenhouse gas emissions come from?', Our World in Data, 18 September 2020, <ourworldindata.org/ghg-emissions-by-sector>.
19   Hannah Ritchie, 'What are the safest and cleanest sources of energy?', Our World in Data, 10 February 2020, <ourworldindata.org/safest-sources-of-energy>.
20   See: Hannah Ritchie and Max Roser, 'Renewable energy', Our World in Data, no date, <ourworldindata.org/renewable-energy>.
21   Hannah Ritchie and Max Roser, 'Nuclear energy', Our World in Data, no date, <ourworldindata.org/nuclear-energy>.
22   'Nuclear and uranium', Australian Greens, no date, <greens.org.au/policies/nuclear-and-uranium>.
23   Martin Rees, *Our Final Hour: A scientist's warning: How terror, error, and environmental disaster threaten humankind's future in this century – on Earth and beyond*, Basic Books, New York, 2003.
24   Ben Goertzel, 'Should humanity build a global AI nanny to delay the singularity until it's better understood?', *Journal of Consciousness Studies*, vol. 19, no. 1–2, 2012, <citeseerx.ist.psu.edu/viewdoc/download?doi=10.1.1.352.3966&rep=rep1&type=pdf>.
25   Eliezer Yudkowsky, 'Artificial intelligence as a positive and negative factor in global risk', in Nick Bostrom and Milan M Ćirković (eds), *Global Catastrophic Risks*, Oxford University Press, Oxford, 2008, p. 333.
26   It's the same with funding. Climate change is the best-funded risk category by far, while AI safety scrapes by with perhaps a few tens of millions of dollars – a paltry sum compared with the tens of billions going into AI R&D. Pandemic prevention also trails climate change investment by hundreds of billions of dollars. For back of the envelope comparisons of funding levels, see Benjamin Todd, 'The case for reducing existential risks', 80,000 Hours, October 2017, <80000hours.org/articles/extinction-risk/>.
27   James Randerson, 'Revealed: The lax laws that could allow assembly of deadly virus DNA', the *Guardian*, 14 June 2006, <www.theguardian.com/world/2006/jun/14/terrorism.topstories3>.
28   Jennifer Couzin-Frankel, 'Poliovirus baked from scratch', *Science*, 11 July 2002, <www.sciencemag.org/news/2002/07/poliovirus-baked-scratch>.
29   Ryan S Noyce, Seth Lederman and David H Evans, 'Construction of an infectious horsepox virus vaccine from chemically synthesized DNA fragments', *PLOS One*, 19 January 2018, <journals.plos.org/plosone/article?id=10.1371/journal.pone.0188453>.
30   For a historical discussion of lab leaks, see: Toby Ord, *The Precipice*, Bloomsbury Publishing, Kindle, 2020.

7   Take x and add AI
1   Ceri Parker, 'Artificial intelligence could be our saviour, according to the CEO of Google', *World Economic Forum*, 24 January 2018,

<www.weforum.org/agenda/2018/01/google-ceo-ai-will-be-bigger-than-electricity-or-fire/>; Catherine Clifford, 'Google CEO: A.I. is more important than fire or electricity', *CNBC*, 1 February 2018, <www.cnbc.com/2018/02/01/google-ceo-sundar-pichai-ai-is-more-important-than-fire-electricity.html>.

2   Benjamin Carlson, 'Quote of the day: Google CEO compares data across millennia', the *Atlantic*, 3 July 2010, <www.theatlantic.com/technology/archive/2010/07/quote-of-the-day-google-ceo-compares-data-across-millennia/344989/>.

3   Bernard Marr, 'How much data do we create every day? The mind-blowing stats everyone should read', *Forbes*, 21 May 2018, <www.forbes.com/sites/bernardmarr/2018/05/21/how-much-data-do-we-create-every-day-the-mind-blowing-stats-everyone-should-read/#666af8af60ba>.

4   Peter H Diamandis, 'Sensors explosion and the rise of IoT', personal website, 9 October 2019, <www.diamandis.com/blog/sensors-and-iot>.

5   'Sizing the prize: What's the real value of AI for your business and how can you capitalise?', PwC, 2017, <www.pwc.com/gx/en/issues/analytics/assets/pwc-ai-analysis-sizing-the-prize-report.pdf>.

6   The American writer and researcher Gwern Branwen has experimented extensively with the current state of the art software in natural-language text generation, OpenAI's GPT-3. He writes: 'GPT-3's samples are not just close to human level: they are creative, witty, deep, meta, and often beautiful'. For a full account, see: 'GPT creative fiction', Gwern Branwen, no date, <www.gwern.net/GPT-3>. Some other great examples of AI-generated art can be found here: Charlie Snell, 'Alien dreams: An emerging art scene', Machine Learning at Berkeley, no date, <ml.berkeley.edu/blog/posts/clip-art/>.

7   Geoff Spencer, 'Much more than a chatbot: China's Xiaoice mixes AI with emotions and wins over millions of fans', Microsoft, 1 November 2018, <news.microsoft.com/apac/features/much-more-than-a-chatbot-chinas-xiaoice-mixes-ai-with-emotions-and-wins-over-millions-of-fans/>; Ralph Haupter, 'Learning to love AI', Microsoft, 14 February 2018, <news.microsoft.com/apac/2018/02/14/learning-love-ai/>.

8   Stuart Russell, *Human Compatible: Artificial intelligence and the problem of control*, Penguin Books London, 2020.

9   Technology for continuous ECG and EEG monitoring already exists and has been applied, with researchers forecasting more pervasive use in the future. See, for example: Sigge Weisdorf et al, 'Ultra-long-term subcutaneous home monitoring of epilepsy – 490 days of EEG from nine patients', *Epilepsia*, vol. 60, no. 11, November 2019, pp. 2204–2214,; <www.ncbi.nlm.nih.gov/pmc/articles/PMC6899579/>; Qing Zhang, Pingping Wang and Yan Liu, 'A real-time wireless wearable electroencephalography system

based on Support Vector Machine for encephalopathy daily monitoring', *International Journal of Distributed Sensor Networks*, 30 May 2018, <https://doi.org/10.1177/1550147718779562>; Shu-Li Guo et al, 'The future of remote ECG monitoring systems, *Journal of Geriatric Cardiology*, vol. 13, no. 6, September 2016; pp. 528–530, <www.ncbi.nlm.nih.gov/pmc/articles/PMC4987424/>; Nilanjan Dey et al, 'Developing residential wireless sensor networks for ECG healthcare monitoring', *IEEE Transactions on Consumer Electronics*, vol. 63, no. 4, November 2017, <ieeexplore.ieee.org/document/8246822>.

10   Eric Topol, *Deep Medicine: How artificial intelligence can make healthcare human again*, Basic Books, Kindle, 2019.

11   I'm evoking a possible future version of Amazon's virtual assistant, Alexa, here. Not because I'm championing, or betting on, Amazon as a frontrunner in this space, but because the name and the product are familiar. That familiarity can help us imaginatively bridge the gap between the present and the future, in a tale that's designed to highlight how AI technology could change all products in the virtual assistant and virtual companion category – and transform us in turn.

12   This bet stacks up with a lot of big-tech predictions, and also the leading AI researcher Stuart Russell's forecast in his 2019 book *Human Compatible*.

13   'Labour force status of families', Australian Bureau of Statistics, 16 October 2020, <www.abs.gov.au/statistics/labour/employment-and-unemployment/labour-force-status-families/latest-release>; Stephanie Kramer, 'U.S. has world's highest rate of children living in single-parent households', Pew Research Center, 12 December 2019, <www.pewresearch.org/fact-tank/2019/12/12/u-s-children-more-likely-than-children-in-other-countries-to-live-with-just-one-parent/>.

14   Susan Harkness, Paul Gregg and Mariña Fernández-Salgad, 'The rise in single-mother families and children's cognitive development: Evidence from three British birth cohorts', Society for Research in Child Development, 20 November 2019, <https://doi.org/10.1111/cdev.13342>.

15   It's true that in the field of behavioural genetics, genes and nonshared environment (that is, environmental influences you don't have in common with others) are thought to account for most of the difference between individual human beings. Nevertheless, shared environment (as in, sharing the same family home, parents and experiences) accounts for a significant portion of environmental influence over some traits. See Robert Plomin, 'Commentary: Why are children in the same family so different? Non-shared environment three decades later', *International Journal of Epidemiology*, vol. 40, no. 3, June 2011, pp. 582–592, <academic.oup.com/ije/article/40/3/582/739694>.

16 Amir Levine and Rachel Heller, *Attached*, Pan Macmillan, London, Kindle, 2012.
17 According to Levine and Heller: 'Just over 50 percent are secure, around 20 percent are anxious, 25 percent are avoidant, and the remaining 3 to 5 percent fall into the fourth, less common category (combination anxious and avoidant)'. See Amir Levine and Rachel Heller, *Attached*.
18 Yisroel Mersky and Wenke Lee, 'The creation and detection of deepfakes: A survey', *ACM Computer Surveys*, vol. 54, no. 1, December 2020, <https://doi.org/10.1145/3425780>.
19 'Weaponised deep fakes – National security and democracy', Australian Strategic Policy Institue, 29 April 2020, <www.aspi.org.au/report/weaponised-deep-fakes>.
20 See: Ana Santos Rutschman, 'Artificial intelligence can now emulate human behaviors – soon it will be dangerously good', the *Conversation*, 5 April 2019, <theconversation.com/artificial-intelligence-can-now-emulate-human-behaviors-soon-it-will-be-dangerously-good-114136>.
21 Ian Sample, 'What are deepfakes – and how can you stop them?', the *Guardian*, 13 January 2020, <www.theguardian.com/technology/2020/jan/13/what-are-deepfakes-and-how-can-you-spot-them; https://www.bbc.com/news/business-51204954>.
22 Nilesh Christopher, 'We've just seen the first use of deepfakes in an Indian election campaign', *Vice*, 18 February 2020, <www.vice.com/en_in/article/jgedjb/the-first-use-of-deepfakes-in-indian-election-by-bjp>.
23 Milan Kundera, *The Unbearable Lightness of Being*, Harper & Row, New York, 1984, p. 100.
24 Kevin Kelly, *The Inevitable: Understanding the 12 technological forces that will shape our future*, Penguin Random House, New York, 2016.
25 Thomas McCabe, 'Singularity and rationality: Eliezer Yudkowsky speaks out', Kurzweil AI, 5 August 2010, <www.kurzweilai.net/singularity-and-rationality-eliezer-yudkowsky-speaks-out>.
26 Elon Musk, Nick Bostrom and Stephen Hawking have publicly expressed this view. Toby Ord provides a quantitative risk estimate in his book *The Precipice*. He estimates the chance of a nuclear war causing an existential catastrophe in the next 100 years at ~1 in 1000. He estimates the risk of unaligned artificial intelligence causing an existential catastrophe in the next 100 years as ~1 in 10.

8   Live forever or die trying
1 See: David Sinclair, *Lifespan: Why we age, and why we don't have to*, Atria Books, New York, 2019. Also, according to Gompertz's Law, for humans, the risk of dying doubles roughly every eight years after the age of 30. See: Kai Kupferschmidt, 'Naked mole rats defy the

biological law of aging', *Science*, 26 January 2018, <www.sciencemag. org/news/2018/01/naked-mole-rats-defy-biological-law-aging>.

2   David Sinclair, *Lifespan*.

3   See, for example: Carlos López-Otín, Maria A Blasco, Linda Partridge, Manuel Serrano and Guido Kroemer, 'The hallmarks of ageing', *Leading Edge Review*, 6 June 2013, <www.cell.com/cell/pdf/S0092-8674(13)00645-4.pdf>.

4   Elie Dolgin, 'Send in the senolytics', *Nature Biotechnology*, vol. 38, 2020, pp. 1371-1377, <www.nature.com/articles/s41587-020-00750-1>.

5   See: Ben Zealley and Aubrey DNJ de Grey, 'Strategies for engineered negligible senesence', *Gerontology*, vol. 59, no. 2, 2012, <https://doi. org/10.1159/000342197>.

6   See, for example: Huber Warner et al, 'Science fact and the SENS agenda', *EMBO Reports*, vol. 6, no. 11, November 2005, pp. 1006-1008, <www.ncbi.nlm.nih.gov/pmc/articles/PMC1371037/>; Bret Weinstein, 'Stop making SENS', *MIT Technology Review*, no date, <www2. technologyreview.com/sens/docs/weinstein.pdf>.

7   See: Editorial, 'Opening the door to treating ageing as a disease', the *Lancet Diabetes & Endocrinogy*, vol. 8, no. 8, 1 August 2018, <www.thelancet.com/journals/landia/article/PIIS2213-8587(18)30214-6/fulltext>; Niz Barzilai, Ana Maria Cuervo and Steve Austad, 'Aging as a biological target for prevention and therapy', *JAMA*, 2 October 2018, <jamanetwork.com/journals/jama/fullarticle/2703112>.

8   Daniel Fabian and Thomas Flatt, 'The evolution of aging', The Nature Education Knowledge Project, 2011, <www.nature.com/scitable/knowledge/library/the-evolution-of-aging-23651151/>.

9   J Graham Ruby, Megan Smith and Rochelle Buffenstein, 'Naked mole-rat mortality rates defy Gompertzian laws by not increasing with age', *eLife*, 24 January 2018, <elifesciences.org/articles/31157>. Note: The authors of the Calico paper, as well as other respondents, have stressed that further research with a larger sample of older mole rats is needed to determine the true lifespan of the naked mole rat, and to more robustly establish whether this species does indeed avoid accelerated ageing in later life. But these initial findings have provided biotech companies and gerontologists with one of a growing number of research avenues worth exploring, in the pursuit of understanding and hacking the human ageing process.

10   'Tiny salamander's huge genome may harbor the secrets of regeneration', *Science Daily*, 28 January 2020, <www.sciencedaily.com/releases/2020/01/200128114638.htm>.

11   Mark Tollis, Amy M Boddy and Carlo C Maley, 'Peto's paradox: How has evolution solved the problem of cancer prevention?', *BMC Biology*, vol. 15, no. 60, 2017, <bmcbiol.biomedcentral.com/articles/10.1186/s12915-017-0401-7>.

12 Mark Tollis et al, 'Peto's paradox'.

13 Gerald Keil, Elizabeth Cummings and João Pedro de Magalhães, 'Being cool: How body temperature influences ageing and longevity', *Biogerontology*, vol. 16, no. 4, pp. 383–397, 2015, .

14 Ray Kurzweil and Terry Grossman, 'Bridges to life' in Gregory M Fahy et al (eds), *The Future of Aging: Pathways to human life-extension*, Springer, Dordrecht, 2010. For yeast, see: Pavlo Kyryakov et al, 'Caloric restriction extends yeast chronological lifespan by altering a pattern of age-related changes in trehalose concentration', *Frontiers in Physiology*, 6 July 2012, <https://doi.org/10.3389/fphys.2012.00256>. For rats, see: Richard Weindruch and Rajinda S Sohal, 'Caloric intake and aging', *New England Journal of Medicine*, vol. 337, no. 14, 2 October 1997, <dx.doi.org/10.1056%2FNEJM199710023371407>. Note that this meta-analysis shows that median lifespan is not improved by dietary restriction in wild-derived mice: Willliam R Swindell, 'Dietary restriction in rats and mice: A meta-analysis and review of the evidence for genotype-dependant effects on lifespan', *Ageing Research Reviews*, vol. 11, no. 2, April 2012, <dx. doi.org/10.1016%2Fj.arr.2011.12.006>. For dogs, see: Dennis F Lawler et al, 'Diet restriction and ageing in the dog,' *British Journal of Nutrition*, vol. 99, 2008, <https://doi.org/10.1017/ s0007114507871686>. For primates, see: Fabian Pifferi et al, 'Promoting healthspan and lifespan with caloric restriction in primates', *Communications Biology*, vol. 2, no. 107, 2019, <www.nature. com/articles/s42003-019-0348-z>.

15 Arthur V Everitt and David G Le Couteur, 'Life extension by calorie restriction in humans', *Annals of the New York Academy of Sciences*, 1 November 2007, <https://doi.org/10.1196/annals.1396.005>.

16 Alessandro Bitto et al, 'Transient rapamycin treatment can increase lifespan and healthspan in middle-aged mice', *eLife*, 23 August 2016, <www.ncbi.nlm.nih.gov/pmc/articles/PMC4996648/>.

17 Joan B Mannick et al, 'mTOR inhibition improves immune function in the elderly', *Science Translational Medicine*, vol. 6, no. 268, 24 December 2014, <stm.sciencemag.org/content/6/268/268ra179>.

18 Patrick Smith et al, 'Regulation of life span by the gut microbiota in the short-lived African turquoise killifish', *eLife*, 22 August 2017, <elifesciences.org/articles/27014>.

19 Steve Horvath, Kavita Singh, Ken Raj et al, 'Reversing age: Dual species measurement of epigenetic age with a single clock', preprint, <www.biorxiv.org/content/10.1101/2020.05.07.082917v1.full.pdf>.

20 Brian K Kennedy et al, 'Geroscience: Linking aging to chronic disease', *Cell*, vol. 159, 6 November 2014, <www.cell.com/cell/pdf/S0092-8674(14)01366-X.pdf>.

21 C Kenyon et al, 'A *C. elegans* mutant that lives twice as long as wild

type', *Nature*, vol. 366, no. 6454, 2 December 1993, pp. 461–464, <pubmed.ncbi.nlm.nih.gov/8247153/>.

22 Nir Barzilai and Toni Robino, *Age Later: Health span, life span, and the new science of longevity*, St. Martin's Publishing Group, New York, 2020, Kindle Edition.

23 Sofiya Milman and Nir Barzilai, 'Dissecting the mechanisms underlying unusually successful human health span and life span', *Cold Spring Harbour Perspectives in Medicine*, vol. 6, no. 1, January 2016, <www.ncbi.nlm.nih.gov/pmc/articles/PMC4691799/>.

24 See, for example: Ceridwen Dovey, 'Can David Sinclair cure old age?', the *Monthly*, September 2018, <www.themonthly.com.au/issue/2018/september/1535724000/ceridwen-dovey/can-david-sinclair-cure-old-age#mtr>.

25 David Sinclair, *Lifespan: Why we age – and why we don't have to*, Atria Books, New York, Kindle, 2019.

26 David Gems, 'Aging: To treat, or not to treat?', *American Scientist*, vol. 99, no. 4, July–August 2011, <www.americanscientist.org/article/aging-to-treat-or-not-to-treat>.

27 Editorial, 'Opening the door to treating ageing as a disease', the *Lancet Diabetes & Endocrinogy, vol.* 8, no. 8, 1 August 2018, <www.thelancet.com/journals/landia/article/PIIS2213-8587(18)30214-6/fulltext>.

28 Alex Zhavoronkov and Bhupinder Bhullar, 'Classifying aging as a disease in the context of ICD-11', *Frontiers in Genetics*, 4 November 2015, <www.ncbi.nlm.nih.gov/pmc/articles/PMC4631811/>.

29 See: Jared M Campbell, Susan M Bellman, Matthew D Stephenson and Karolina Lisy, 'Metformin reduces all-cause mortality and diseases of ageing independent of its effect on diabetes control: A systematic review and meta-analysis', *Ageing Research Reviews*, vol. 40, November 2017, pp. 31–44, <pubmed.ncbi.nlm.nih.gov/28802803/>; CA Bannister et al, 'Can people with type 2 diabetes live longer than those without? A comparison of mortality in people initiated with metformin or sulphonylurea monotherapy and matched, non-diabetic controls', *Diabetes, Obesity and Metabolism*, vol. 16, no. 11, November 2014, pp. 1165–1173, <pubmed.ncbi.nlm.nih.gov/25041462/>.

30 'Targeting the biology of aging. Ushering a new era of interventions', American Federation for Aging Research, no date, <www.afar.org/tame-trial>.

31 See: Hannah Ritchie, 'Cancer death rates are falling; five-year survival rates are rising', Our World in Data, 4 February 2019, <ourworldindata.org/cancer-death-rates-are-falling-five-year-survival-rates-are-rising>; Stacy Simon, 'Facts & figures 2020 reports largest one-year drop in cancer mortality', American Cancer Society, 8 January 2020, <www.cancer.org/latest-news/facts-and-figures-2020.html>; 'Cancer in Australia statistics', Australian Government Cancer

Australia, 2021, <www.canceraustralia.gov.au/research/data-and-statistics/cancer-statistics>.

32   Andrew Powaleny, 'ICYMI: New study shows medicines advance life expectancy for HIV patients', PhRMA, 15 May 2017, <https://catalyst.phrma.org/icymi-new-study-shows-medicines-advance-life-expectancy-for-hiv-patients>.

33   Max Roser, Hannah Ritchie and Bernadeta Dadonaite, 'Child and infant mortality', Our World in Data, November 2019, <ourworldindata.org/child-mortality>.

34   Hannah Devlin, '60-year-old maths problem partly solved by amateur', the *Guardian*, 5 May 2018, <www.theguardian.com/science/2018/may/04/60-year-old-maths-problem-partly-solved-by-amateur>.

35   See: Ceridwen Dovey, 'Can David Sinclair cure old age?'; Cynthia Kenyon, 'The first long-lived mutants: Discovery of the insulin/IGF-1 pathway for ageing', *Philosophical Transactions of the Royal Society Biological Sciences*, vol. 366, no. 1561, pp. 9–16, 12 January 2011, <www.ncbi.nlm.nih.gov/pmc/articles/PMC3001308/>. The journalist and author Brian Alexander also reported that in the 1990s, biologists began talking more about altering the human genome and radically transforming human biology, though 'most who did had a habit of reaching across desks to flick off tape recorders before using words like life span extension, immortality, or human enhancement'. For further discussion, see Brian Alexander, *Rapture: How biotech became the new religion*, Perseus Books, New York, 2003, p. 224.

36   For a brief discussion, see: Toby Ord, *The Precipice: Existential risk and the future of humanity*, Hachette, London, 2020.

37   MC Liu, GR Oxnard et al, 'Sensitive and specific multi-cancer detection and localization using methylation signatures in cell-free DNA', *Annals of Oncology*, 30 March 2020, <www.annalsofoncology.org/article/S0923-7534(20)36058-0/fulltext>.

38   Susie Neilson, Andrew Dunn and Aria Bendix, 'Moderna's groundbreaking coronavirus vaccine was designed in just 2 days', *Business Insider*, 27 November 2020, <www.businessinsider.com.au/moderna-designed-coronavirus-vaccine-in-2-days-2020-11?r=US&IR=T>.

39   See: Robert Freitas, Nanomedicine series, Landes Bioscience, London, 1999.

40   'Ageing and health', World Health Organization, 4 October 2021, <www.who.int/news-room/fact-sheets/detail/ageing-and-health>.

41   Hannah Ritchie, 'The world population is changing: For the first time there are more people over 64 than children younger than 5', Our World in Data, 23 May 2019, <ourworldindata.org/population-aged-65-outnumber-children>.

42   See: 'Population ages 0–14 (% of total population) – India, United States', World Bank, 2019 <data.worldbank.org/indicator/SP.POP.0014.TO.ZS?locations=IN-US>.

43 Data from India's National Family Health Survey (NFHS-5),
2019–2021, reproduced in: the *Economist*, 'India's population will start
to shrink sooner than expected', 4 December 2021, <www.economist.
com/asia/2021/12/02/indias-population-will-start-to-shrink-sooner-
than-expected>.
44 'Population ages 0–14 (% of total population)', World Bank, 2019,
<data.worldbank.org/indicator/SP.POP.0014.TO.ZS>.
45 Sebastian Thuault, 'Reflections on aging research from within the
National Institute on Aging', *Nature Aging*, vol. 1, pp. 14–18, 2021,
<www.nature.com/articles/s43587-020-00009-z>.
46 Nir Barzilai, TEDMED 2016, 'Can we grow older without growing
sicker?' <www.youtube.com/watch?v=MGKB9AdPmwc>.

9 The post-work society

1 Ava would graduate from a four-year undergraduate degree in 2044.
This is too far ahead for us to make sound predictions about what
kinds of jobs will exist and what their exact nature will be. But
the potential for automation over the next 22 years is profound,
particularly in light of the strong possibility of continued exponential
– or at least rapid – growth in the development of information
technologies. In 2013, the economist Carl Benedikt Frey and the
machine learning researcher Michael A Osborne estimated that
47 per cent of US jobs were at high risk of automation. They were
careful to state that they made no predictions about how many
jobs would be lost, or when, just what the potential was. See: Carl
Benedikt Frey and Michael A Osbone, 'The future of employment:
How susceptible are jobs to automation?', *Technological Forecasting
and Social Change*, vol. 114, January 2017, <https://doi.org/10.1016/j.
techfore.2016.08.019> (originally published as a working paper by
the Oxford Martin school in 2013). In 2017, McKinsey found that
'about 60 percent of all occupations have at least 30 percent of activities
that are technically automatable, based on currently demonstrated
technologies'. See: James Manyika, 'Technology, jobs, and the future
of work,' McKinsey 24 May 2017, <www.mckinsey.com/featured-
insights/employment-and-growth/technology-jobs-and-the-future-of-
work#>.
2 This has been convincingly argued by Federico Pistono, among others,
in *Robots Will Steal Your Job But That's OK: How to survive the economic
collapse and be happy*, CreateSpace, Kindle, 2014.
3 See: Kathryn Diss and Inga Ting, 'A tale of two cities revelas vast gulf
in housing affordability,' ABC News, 5 December 2017, <www.abc.net.
au/news/2017-12-05/how-long-does-it-take-to-save-a-home-deposit-
in-australia/9225272?nw=0>; Richard Cooke, 'The boomer supremacy',
the *Monthly*, March 2016, <www.themonthly.com.au/issue/2016/
march/1456750800/richard-cooke/boomer-supremacy#mtr>.

4　David Chau, 'Australian property "severely unaffordable", Sydney crowned second least affordable market', ABC News, 22 January 2018, <www.abc.net.au/news/2018-01-22/australian-housing-unaffordability-experts-disagree-on-extent/9349796>.

5　McKinsey Global Institute, 'What can history teach us about technology and jobs?', podcast transcript, McKinsey, 16 February 2018, <www.mckinsey.com/featured-insights/future-of-work/what-can-history-teach-us-about-technology-and-jobs>.

6　McKinsey Global Institute, 'Jobs lost, jobs gained: Workforce transitions in a time of automation', McKinsey, December 2017, <www.mckinsey.com/featured-insights/future-of-work/jobs-lost-jobs-gained-what-the-future-of-work-will-mean-for-jobs-skills-and-wages>.

7　Stuart Russell, *Human Compatible: AI and the problem of control*, Penguin Random House, Kindle, 2019.

8　See: 'Artificial Intelligence: Implications for China', McKinsey, April 2017, <www.mckinsey.com/featured-insights/china/artificial-intelligence-implications-for-china>.

9　McKinsey Global Institute, 'Jobs lost, jobs gained'.

10　See: National Safety Council Injury Facts, 'Odds of dying', National Safety Council, no date, <injuryfacts.nsc.org/all-injuries/preventable-death-overview/odds-of-dying/>; Katherine Ellen Foley, 'The death rate for opioid use has surpassed car crashes in the US', *Quartz*, 15 January 2019, <qz.com/1524186/the-death-rate-for-opioid-use-has-surpassed-car-crashes-in-the-us/>.

11　Andrew Yang, *The War on Normal People: The truth about America's disappearing jobs and why universal basic income is our future*, Hachette Books, Kindle, 2018.

12　Andrew Yang, *The War on Normal People*.

13　Andrew Yang, *The War on Normal People*.

14　Anton Cheremukhin, 'Middle-skill jobs lost in U.S. labor market polarisation', *Dallas Fed 9.5*, May 2014, <www.dallasfed.org/~/media/documents/research/eclett/2014/el1405.pdf>.

15　Guido Matias Cortes, Nir Jaimovich and Henry E Siu, 'The "end of men" and rise of women in the high-skilled labor market', Working Paper, *National Bureau of Economic Research*, February 2018, revised November 2018, <www.nber.org/system/files/working_papers/w24274/w24274.pdf>.

16　Jon Birger, *Date-onomics: How dating became a lopsided numbers game*, Workman Publishing Company, Kindle, 2015.

17　Albert Esteve et al, 'The end of hypergamy: Global trends and implications', *Population and Development Review*, 21 November 2016, <https://doi.org/10.1111/padr.12012>.

18　David Graeber, *Bullshit Jobs*, Penguin Books, Kindle, 2018.

19 Elizabeth Barlow, 'The *New York Magazine* environmental teach-in', *New York Magazine*, 30 March, 1970, <books.google.com. au/books?id=cccDAAAAMBAJ&printsec=frontcover&redir_ esc=y#v=onepage&q&f=false>.

20 See, for example: Hunt Allcott et al, 'The welfare effects of social media', *American Economic Review*, vol. 110, no. 3, March 2020, <10.1257/aer.20190658>; Jean M Twenge et al, 'Increases in depressive symptoms, suicide-related outcomes, and suicide rates among U.S. adolescents after 2010 and links to increased new media screen time', *Clinical Psychological Science*, vol. 6, no. 1, 14 November 2017, <https://doi.org/10.1177/2167702617723376>; Elroy Boers et al, 'Association of screen time and depression in adolescence', *JAMA Pediatrics*, 15 July 2019, <doi:10.1001/jamapediatrics.2019.1759>; Ariel Shensa et al, 'Problematic social media use and depressive symptoms among U.S. young adults: A nationally-representative study', *Social Science & Medicine*, vol. 182, June 2017, <https://doi.org/10.1016/j.socscimed.2017.03.061>; Alyssa N Saiphoo, Lilach Dahoah Halevi and Zahra Vahedi, 'Social networking site use and self-esteem: A meta-analytic review', *Personality and Individual Differences*, vol. 153, 15 January 2020, <www.sciencedirect.com/science/article/abs/pii/S0191886919305719>.

10 A generation of kidults

1 This summary of characteristics that delineate Gen Z from previous generations is drawn from Jean Twenge's work in *iGen: Why today's super-connected kids are growing up less rebellious, more tolerant, less happy – and completely unprepared for adulthood – and what that means for the rest of us*, Atria Books, Kindle, 2017.
My claim is not that all members of Gen Z exhibit these characteristics, it's these characteristics are more common among (and in some cases, unique to) Gen Z, relative to previous generations at the same age.

2 Though the increases aren't huge, suicide rates increased for all age groups in the United States between 2009 and 2018. See: Suicide Prevention Resource Center, 'Suicide rates by age, United States 2010–2019', <sprc.org/scope/age>. Suicide rates have remained roughly constant in Australia for 15- to 17-year-olds, but have steadily increased for 18- to 24-year-olds since 2012. See: Australian Institute of Health and Welfare, 'Deaths by suicide among young people', 14 October 2021, v.10.0, <www.aihw.gov.au/suicide-self-harm-monitoring/data/populations-age-groups/suicide-among-young-people>.

3 Jean Twenge, *iGen*.

4 Jean Twenge, *iGen*.

5   Kim Parker and Ruth Igielnik, 'On the cusp of adulthood and
     facing an uncertain future: What we know about Gen Z so far',
     Pew Research Center, 14 May 2020, <www.pewresearch.org/social-
     trends/2020/05/14/on-the-cusp-of-adulthood-and-facing-an-
     uncertain-future-what-we-know-about-gen-z-so-far-2/>.

6   Christopher Ingraham, 'The share of Americans not having sex has
     reached a record high', *Washington Post*, 29 March 2019,.

7   Eric Mack, 'A quarter of Japanese adults under 40 are virgins, and the
     number is increasing', *Forbes*, 7 April 2019, <www.forbes.com/sites/
     ericmack/2019/04/07/a-quarter-of-japanese-adults-under-40-are-
     virgins-and-the-number-is-increasing/?sh=255cc3c67e4d>.

8   A helpful summary of some of this research can be found in: Bradford
     Tuckfield, 'Attraction inequality and the dating economy', *Quillette*,
     12 March 2019, <quillette.com/2019/03/12/attraction-inequality-and-
     the-dating-economy/>. For additional material, see: Dan Kopf, 'These
     statistics show why it's so hard to be an average man on dating apps',
     QZ, 15 August 2017, <qz.com/1051462/these-statistics-show-why-
     its-so-hard-to-be-an-average-man-on-dating-apps/>; Worst-Online-
     Dater, 'Tinder Experiments II: Guys, unless you are really hot, you are
     probably better off not wasting your time on Tinder – a quantitative
     socio-economic study', *Medium*, 25 March 2015, <medium.com/
     @worstonlinedater/tinder-experiments-ii-guys-unless-you-are-really-
     hot-you-are-probably-better-off-not-wasting-your-2ddf370a6e9a>.

9   Jon Birger, *Date-onomics: How dating became a lopsided numbers game*,
     Workman Publishing Company, Kindle, 2015.

10   Jon Birger, *Date-onomics*.

11   Briony Smith, 'Why being single sucks: What no one wants to talk
     about', *Fashion*, 29 December 2018, <www.flare.com/tv-movies/why-
     being-single-sucks-what-no-one-wants-to-talk-about/>.

12   This comment was posted on the subreddit IncelsWithoutHate in
     2020. This community has subsequently banned from Reddit for
     'violating Reddit's rule against promoting hate'. I think it's very
     important to quote this now 'disappeared' content, as this post
     was an anguished cry – from one of many human beings among a
     community that's suffering and that appears to be at a loss to know
     how to heal. We cannot understand this phenomenon, or this brand
     of anguish, if we silence it and deem every expression of it a danger to
     civilisation. This post was previously available at: <www.reddit.com/r/
     IncelsWithoutHate/comments/iisuli/im_an_incel_at_28/>.

13   For an introductory summary, see: Graeme Wood, 'The next decade
     could be even worse', the *Atlantic*, December 2020,.

14 Yvonne Roberts, 'Millennials are struggling. Is it the fault of the baby boomers?', the *Guardian*, 29 April 2018, <www.theguardian.com/society/2018/apr/29/millennials-struggling-is-it-fault-of-baby-boomers-intergenerational-fairness>.

15 Yvonne Roberts, 'Millennials are struggling'.

16 See: Elise Bohan, *A History of Transhumanism*, doctoral thesis submitted for examination, Macquarie University, 2018. Macquarie University Research Online, <hdl.handle.net/1959.14/1271515>.

17 Hank Pellissier, 'Transgender and Transhuman – the alliance, the complaints and the future', *Institute for Ethics and Emerging Technologies (IEET)*, 15 June 2012, <web.archive.org/web/20120618025654/http://ieet.org/index.php/IEET/more/pellissier20120615>.

18 Jan Morris, *Conundrum*, Faber and Faber, 1974, pp. 44–45.

19 For the period spanning 2008–2018, Sweden's Board of Health and Welfare reported a 1500 per cent increase of gender dysphoria diagnoses in girls aged 13–17 years. See: Richard Orange, 'Teenage transgender row splits Sweden as dysphoria diagnoses soar by 1,500%', the *Guardian*, 22 February 2020, <www.theguardian.com/society/2020/feb/22/ssweden-teenage-transgender-row-dysphoria-diagnoses-soar>. In the United States, the prevalence of adolescent gender dysphoria increased by 1000 per cent in the 2010s, while a 4000 per cent increase was seen in Britain. See: Abigail Shrier, *Irreversible Damage: The transgender craze seducing our daughters*, Regnery Publishing, Kindle, 2020.

20 Risks of taking puberty blockers, and undergoing hormone therapy and gender reassignment procedures include: bone density loss, heart damage, vaginal wall atrophy in natal females, changes in sexual response and sensation, and permanent sterility. See: Micol S Rothman and Sean J Iwamoto, 'Bone health in the transgender population', *Clinical Reviews in Bone and Mineral Metabolism*, vol. 17, 2019, pp. 77–85, <https://doi.org/10.1007/s12018-019-09261-3>; Grant Ferguson et al, 'Gender dysphoria: Puberty blockers and loss of bone mineral density', *British Medical Journal*, 4 December 2019, <https://doi.org/10.1136/bmj.l6471>; Michael S Irwig, 'Cardiovascular health in transgender people', *Reviews in Endocrine and Metabolic Disorders*, vol. 19, 2018, pp. 243–251, <https://doi.org/10.1007/s11154-018-9454-3>; Maurizio Baldassarre et al, 'Effects of long-term high dose testosterone administration on vaginal epithelium structure and estrogen receptor-α and -β expression of young women', *International Journal of Impotence Research*, vol. 25, 2013, pp. 172–177, <https://doi.org/10.1038/ijir.2013.9>; Philip J Cheng, 'Fertility concerns of the transgender patient', *Translational Andrology and Urology*, vol. 8, no. 3, 2019, pp. 209–218, <dx.doi.org/10.21037%2Ftau.2019.05.09>.

11 The future of sex

1 Benjamin Haas, 'Chinese man "marries" robot he built himself', the *Guardian*, 4 April 2017, <www.theguardian.com/world/2017/apr/04/chinese-man-marries-robot-built-himself>.

2 Viola Zhou, 'Inkstone index: China's gender imbalance', Inkstone, 18 January 2019, <www.inkstonenews.com/society/inkstone-index-china-has-worlds-largest-gender-imbalance/article/3000518>.

3 Mei Fong, 'Sex dolls are replacing China's missing women', *Foreign Policy*, 28 September 2017, <foreignpolicy.com/2017/09/28/sex-dolls-are-replacing-chinas-missing-women-demographics/>.

4 Quanbao Jiang and Jesús J Sánchez-Barricarte, 'The predicament of bare branches' sexuality', *Electronic Journal of Human Sexuality*, vol. 15, 18 September 2012, <www.ejhs.org/volume15/Bare.html>.

5 Quanbao Jiang and Jesus J Sanchez-Barricarte, 'The predicament of bare branches' sexuality'.

6 Mitchell Langcaster-James and Gillian R Bentley, 'Beyond the sex doll: Post-human companionship and the rise of the "allodoll"', *Robotics*, vol. 7, no. 4, 2018, p. 62, <www.mdpi.com/2218-6581/7/4/62>.

7 'Realdoll testimonials', RealDoll, no date, <www.realdoll.com/testimonials/>.

8 Felix Allen, 'My sex doll is so much better than my real wife', the *New York Post*, 30 June 2017, <nypost.com/2017/06/30/i-love-my-sex-doll-because-she-never-grumbles/>.

9 Laurie Segall, 'Mostly human: I love you, bot', CNN, 8 March 2017, <edition.cnn.com/videos/cnnmoney/2017/03/08/mostly-human-i-love-you-bot.cnnmoney>.

10 Kathleen Richardson, 'Robots and ethics: The future of sex', TEDx Talk, 14 June 2016, <www.youtube.com/watch?v=YaMiH93-iPE>.

11 'Our story', Campaign Against Sex Robots, no date, <campaignagainstsexrobots.org/our-story/>.

12 'Ethics of robots', Campaign Against Sex Robots, no date, <campaignagainstsexrobots.org/ethics-of-robots/>.

13 RumFuelledRedHead90 reply to u/throwaway190596, '[Serious] women of Reddit, sex dolls are becoming cheaper and more realistic looking, what are your thoughts on this topic?', Reddit thread, 2017, <www.reddit.com/r/AskReddit/comments/7pfg7t/serious_women_of_reddit_sex_dolls_are_becoming/>.

14 Diana Fleischman, 'Uncanny vulvas', Dianaverse blog, 30 October 2020, <dianaverse.com/2020/10/30/uncanny-vulvas/>.

15 Diana Fleischman, 'Uncanny vulvas'.

12 The end of having babies

1 'Total fertility rate', World Population Review, 2021, <worldpopulationreview.com/countries/total-fertility-rate/>.

2 Max Roser, 'Fertility rate', Our World in Data, 2 December 2017,

<ourworldindata.org/fertility-rate>.

3   World Population Review, 'Total fertility rate'.
4   Max Roser, 'When will the world reach "peak child"?', Our World in Data, 8 February 2018, <ourworldindata.org/peak-child>.
5   Seth Wynes and Kimberly A Nicholas, 'The climate mitigation gap: Education and government recommendations miss the most effective individual actions', *Environmental Research Letters*, vol. 12, no. 7, 12 July 2017, <iopscience.iop.org/article/10.1088/1748-9326/aa7541>.
6   David Sinclair, *Lifespan: Why we age – and why we don't have to*, Atria Books, New York, Kindle, 2019.
7   Sonia M Suter, 'In vitro gametogenesis: Just another way to have a baby?', *Journal of Law and the Biosciences*, vol. 3, no. 1, December 2015, <https://doi.org/10.1093/jlb/lsv057>; I Glenn Cohen George, Q Daley and Eli Y Adashi, 'Disruptive reproductive technologies', *Science Translational Medicine*, vol. 9, no. 372, 11 January 2017, <https://doi.org/10.1126/scitranslmed.aag2959>.
8   Nick Bostrom, *Superintelligence: Paths, dangers, strategies*, Oxford University Press, Kindle, Oxford, 2016.
9   JBS Haldane, *Daedalus: Or science and the future*, EP Dutton and Company, New York, 1923, pp. 65, 68.
10   Emily A Partridge et al, 'An extra-uterine system to physiologically support the extreme premature lamb', *Nature Communications*, vol. 8, 25 April 2017, <www.nature.com/articles/ncomms15112?utm_source=commission_junction&utm_medium=affiliate>.
11   Nicola Davis, 'Artificial womb: Dutch researchers given €2.9m to develop prototype', the *Guardian*, 8 October 2019, <www.theguardian.com/society/2019/oct/08/artificial-womb-dutch-researchers-given-29m-to-develop-prototype>.
12   Gina Kolata, 'Scientists grow mouse embryos in a mechanical womb', *New York Times*, 17 March 2021, <www.nytimes.com/2021/03/17/health/mice-artificial-uterus.html?action=click&module=Top%20Stories&pgtype=Homepage>.
13   Mike Bird, 'The Covid baby bust could reverberate for decades', the *Wall Street Journal*, 5 March 2021, <www.wsj.com/articles/the-covid-baby-bust-could-reverberate-for-decades-11614962945?reflink=desktopwebshare_twitter>.
14   Robert Ettinger, *Man into Superman: The startling potential of human evolution – and how to be part of it* (plus additional comments by others 'Developments in transhumanism 1972–2005'), edited by Charles Tandy, Ria University Press, Palo Alto, 2005.

Postscript: The start of something new
1   John Stuart Mill, *Utilitarianism*, Parker, Son and Bourn, London, 1863, p. 14.